THINK CODE CREATE

A Python Programming Journey

Monte North

Second Edition

Think, Code, Create

A Python Programming Journey

Second Edition

Monte North
BEd.(Sec.) • BSc. • MIT.

2026

First published 2025

Second Edition published 2026

National Library of Australia Cataloguing-in-Publication entry:

North, Monte, author.
Think, Code, Create: A Python Programming Journey / Monte North.
Second edition.
ISBN: 978-1-7640698-9-2
Subjects: Python (Computer program language). Computer programming.

Cover design: Majella T North

Typeset in Palatino Linotype and Consolas

Printed in Australia

To every curious mind that has ever opened a terminal and wondered what is possible.
And to the teachers who refuse to let the question stay unanswered.

About the Author

> **Monte North BEd.(Sec.) BSc. MIT.** is a Masters-qualified educator and ICT professional whose career spans more than three decades across network engineering, cybersecurity, applied mathematics, and secondary education.

Monte holds a Master of Information Technology from Queensland University of Technology with Distinction, specialising in Network Management and Cybersecurity, alongside a Bachelor of Science in Mathematics and Computer Science, a Bachelor of Education (Secondary) — both from the University of Queensland.

His industry career spans more than three decades: from network engineering roles at public and private companies across Australia until forming his own company, NCS Professionals — which involved cybersecurity consulting, penetration testing, security risk assessments, and network architecture design for corporate clients. He is proficient in Python, C#, Java, C, C++, PHP, and MySQL, and has hands-on experience across Linux, Windows, Cisco enterprise environments, and multiple intrusion detection platforms.

Alongside his industry work, Monte has had a long teaching career with Education Queensland, delivering Digital Technology, IT, Mathematics, and Physics across a range of secondary schools. Think, Code, Create was born from that classroom experience — written by someone who has built production networks and stood at the front of a room explaining recursion to teenagers. That combination is what makes it distinctive.

Preface to the Second Edition

The first edition of Think, Code, Create set out to do something that many programming textbooks avoid: treat the reader as a capable adult, regardless of their starting point. It was written for a broad audience — curious beginners, career changers, professionals looking to add Python to their toolkit, and secondary students encountering programming in a formal context for the first time. The response confirmed that this approach was the right one. Readers found the book rigorous without being intimidating, and practical without being shallow.

This second edition keeps that foundation intact. Every change made here has been guided by a single question: does this make the book more useful for the reader it is actually speaking to? Where the answer was yes, the change was made. Where it was not, the original text was left alone.

What Has Changed

The most substantial updates appear in the chapter content and worked examples. Throughout the first edition, examples occasionally leaned on school-based contexts — classroom scores, quiz questions, fictional student records — that suited a narrow audience but underserved the majority. In this edition, those examples have been replaced with real-world scenarios drawn from professional practice: log file parsing, network configuration, financial modelling, data pipeline processing, and cybersecurity tooling. The code is no more complex than before — it simply sits in contexts that feel worth caring about.

The appendices have been comprehensively rewritten. Each section now includes a minimum of two worked examples demonstrating the same concept under different parameters, so that readers can see not just how a feature works but how its behaviour varies. Explanations have been tightened and, in several sections, significantly expanded to address the gaps that emerged from reader feedback.

The chapter structure and Bloom's Taxonomy review exercise format have been retained. These were consistently praised in feedback as a genuine aid to consolidation, and the exercise prompts have been updated to match the revised professional framing of the main content. No school-based exercises remain.

Australian spelling has been applied consistently throughout. Coding conventions have been brought into alignment with current PEP 8 standards, including updated guidance on naming conventions, type checking with isinstance(), and the if __name__ == '__main__': pattern for script entry points.

What Has Not Changed

The structure of the book — fifteen chapters moving from foundational concepts through to advanced function patterns, data structures, and the Python ecosystem — remains as it was. The pedagogical approach remains the same: concepts are introduced clearly, illustrated with purposeful examples, and then consolidated through exercises that require the reader to actually think. The voice of the book is unchanged.

A Note on the Cover

This edition carries a new cover, selected following structured testing with a representative reader panel. The design was chosen because it communicates what the book actually is: a technical resource that takes its subject seriously, written for people who do too.

Monte North
Brisbane, Queensland
2026

How to Use This Book

This book is designed to be read sequentially, but it does not require you to be. If you already have programming experience in another language, you may find that you can move quickly through Chapters 1 through 4 and accelerate from Chapter 5 onwards. If you are approaching programming for the first time, working chapter by chapter is strongly recommended — each chapter assumes familiarity with the concepts introduced in those before it.

Chapter Structure

Each chapter follows a consistent format. An introductory section sets out what the chapter covers and why it matters. Numbered sections then develop the topic in a logical sequence, with worked code examples throughout. A chapter summary recaps the key ideas, and the chapter closes with a set of review exercises.

Review Exercises

The review exercises at the end of each chapter are structured according to Bloom's Taxonomy, progressing from recall and comprehension through to application, analysis, evaluation, and creation. The early exercises in each set are deliberately straightforward — they are checking that the fundamentals have been absorbed. The later exercises require genuine problem-solving. Do not skip them.

Code Examples

All code in this book has been written and tested with Python 3.10 and above. If you are running an earlier version of Python 3, the vast majority of examples will still run correctly, but a small number of examples use match statements and structural pattern matching introduced in Python 3.10. These are clearly labelled.

Code examples use meaningful, professional contexts wherever possible. Variables have descriptive names. Functions do one thing well. The style follows PEP 8 throughout. Reading code with good style is as valuable as reading the explanations — the examples are teaching habits, not just syntax.

The Appendices

The appendices are reference material, not prerequisite reading. They cover string methods, built-in functions, common data structure operations, and standard library modules in depth, with multiple worked examples per section. Consult them when you encounter a method or function in the chapters and want a fuller account of its behaviour.

A Note on Running the Code

Recommended setup

Python: Download from python.org. Use version 3.10 or later.

Editor: Visual Studio Code with the Python extension is strongly recommended.

Environment: Use a virtual environment (venv) for any project that installs third-party packages.

Terminal: Familiarise yourself with your operating system's terminal. Chapter 2 covers this.

The best way to use this book is with a terminal open alongside it. Type the examples. Modify them. Break them deliberately and observe what happens. Programming is a practical discipline — reading about it is useful, but it is not the same as doing it.

TABLE OF CONTENTS

Think, Code, Create – A Python Programming Journey

CHAPTER 1 Introduction to Programming

Think, Code, Create – A Python Programming Journey

> This chapter establishes the conceptual foundation on which everything else in this book rests. Before writing useful programs, it helps to understand what programming actually is, how different languages approach the same problems, and why Python has become the language of choice across such a wide range of industries and disciplines. Whether you are approaching programming for the first time or arriving with experience in another language, this chapter will orient you within the broader landscape of software development.
>
> We begin with the nature of programming itself, then explore the major paradigms that shape how developers think about code. From there we examine how languages are executed, compare Python with several widely used alternatives, and survey the real-world domains where Python is applied every day. The chapter closes with an overview of the tools and libraries that make up Python's ecosystem — the practical environment in which you will be working throughout this book.

1.1 What is Programming?

Programming is the practice of writing precise instructions that a computer can follow to perform a task. At its most fundamental level, a program is a sequence of operations: read some data, transform it according to defined rules, and produce a result. The range of tasks that can be expressed this way is extraordinary — from calculating a loan repayment schedule to coordinating the movement of autonomous vehicles, from filtering spam email to rendering three-dimensional graphics in real time.

A computer is an extremely capable machine, but it has no capacity for inference or assumption. It does exactly what it is instructed to do, nothing more. This is what makes programming both powerful and demanding: the programmer must account for every case, every possible input, and every path through the logic. When those instructions are clear and complete, the result is software that is reliable and fast. When they are not, the result is bugs.

Programs are written in programming languages — formal systems with precise grammar rules that sit between human thought and machine execution. When you write code in Python, you are working in a high-level language: one designed to be readable and expressive to humans, which is then translated into low-level instructions the hardware can process.

A practical example

The following program reads a list of daily temperatures and reports the average, maximum, and the number of days that exceeded a defined threshold:

Example 1 — Introduction

Analysing a week of temperature readings from a weather station.

```
temperatures = [22.1, 19.8, 31.4, 28.7, 17.3, 33.2, 25.0]
threshold = 30.0

average    = sum(temperatures) / len(temperatures)
maximum    = max(temperatures)
days_over  = sum(1 for t in temperatures if t > threshold)

print(f"Average temperature: {average:.1f}C")
print(f"Maximum temperature: {maximum:.1f}C")
print(f"Days exceeding {threshold}C: {days_over}")
```

Each line performs a specific, unambiguous operation. If the list were empty, the division on the fourth line would raise an error because dividing by zero is undefined. That edge case must be handled explicitly if it is a real possibility. This illustrates a key reality of programming: thinking carefully about what can go wrong is as important as writing the logic for what should go right.

Errors are a normal part of the process. A syntax error occurs when the code violates the grammatical rules of the language — a missing bracket, a misspelled keyword, incorrect indentation. Python will detect these before executing anything and report the location. A logic error is subtler: the code runs without complaint but produces the wrong result because the instructions, while valid, do not correctly capture the intended behaviour. Learning to read error messages, trace execution, and isolate the source of a problem — debugging — is a skill that develops with practice and is central to professional programming.

1.2 Overview of Programming Paradigms

A programming paradigm is a fundamental style or approach to structuring code. Paradigms are not features of specific languages — they are philosophies about how to decompose problems and organise solutions. Most modern languages, including Python, support multiple paradigms, giving developers the flexibility to choose the approach that best fits the problem at hand.

The three paradigms most relevant to Python programming are procedural, object-oriented, and functional.

Procedural Programming

Procedural programming structures a program as a sequence of instructions executed in order, grouped into functions that can be called as needed. It is the most direct expression of how a computer operates — step by step, with branches and loops to vary the path. For many practical tasks, procedural code is the clearest and most natural approach.

Example 1 — Introduction

A procedural pipeline that filters sensor readings and computes a clean average.

```
def filter_readings(readings, min_val, max_val):
    return [r for r in readings if min_val <= r <= max_val]
```

```python
def compute_average(readings):
    if not readings:
        return None
    return sum(readings) / len(readings)

raw_data = [4.2, 999.9, 3.8, 4.5, -1.0, 4.1, 4.7]
clean    = filter_readings(raw_data, 0.0, 10.0)
avg      = compute_average(clean)
print(f"Valid readings: {clean}")
print(f"Average: {avg:.2f}")
```

Each function does one thing. The main flow is easy to read from top to bottom. Procedural programming is well-suited to scripts, automation tasks, and any problem where the logic is largely linear.

Object-Oriented Programming (OOP)

Object-oriented programming organises code around objects — self-contained units that combine data (attributes) and behaviour (methods). Objects are created from templates called classes. OOP is particularly effective when a program models entities that have persistent state and interact with each other — accounts in a banking system, devices in a network, orders in an e-commerce platform.

Example 2 — Variation

A BankAccount class that encapsulates balance management and enforces valid operations.

```python
class BankAccount:
    def __init__(self, owner, initial_balance=0.0):
        self.owner   = owner
        self.balance = initial_balance

    def deposit(self, amount):
        if amount > 0:
            self.balance += amount
            print(f"Deposited ${amount:.2f}. Balance: ${self.balance:.2f}")

    def withdraw(self, amount):
        if amount > self.balance:
            print("Insufficient funds.")
        else:
            self.balance -= amount
            print(f"Withdrew ${amount:.2f}. Balance: ${self.balance:.2f}")

account = BankAccount("Sarah Chen", 1000.00)
account.deposit(500.00)
account.withdraw(200.00)
```

Each BankAccount object maintains its own state independently. OOP's key strengths — encapsulation, inheritance, and polymorphism — become especially valuable as systems grow in complexity.

Functional Programming

Functional programming treats computation as the evaluation of mathematical functions, emphasising immutability and the avoidance of side effects. Rather than modifying data in place, functional code transforms data by passing it through a series of functions, each producing a new value without altering the original.

Example 3 — Real-world application

A functional pipeline that filters and totals financial transactions without modifying any original data.

```
from functools import reduce

transactions = [120.0, -45.0, 300.0, -180.0, 95.0, -20.0]

credits = list(filter(lambda t: t > 0, transactions))
total   = reduce(lambda acc, t: acc + t, credits, 0.0)

print(f"Credits: {credits}")
print(f"Total credits: ${total:.2f}")
```

No data was modified in place. Each operation produced a new value. The paradigms are not mutually exclusive — Python programs commonly blend all three. The skill is knowing which tool fits the problem.

1.3 Compiled vs Interpreted Languages

When you write code in a programming language, it must eventually be converted into instructions the processor can execute. The mechanism by which this conversion happens defines one of the most important distinctions in the field: whether a language is compiled or interpreted. Understanding this distinction helps explain differences in performance, development workflow, portability, and error behaviour.

Compiled Languages

A compiled language requires the source code to be translated into machine code before the program can run. This translation is performed by a compiler, which processes the entire source in one pass, checks it for errors, applies optimisations, and produces an executable file. Once compiled, the program can be run repeatedly at close to the maximum speed the hardware supports. C and C++ are the canonical compiled languages.

Key characteristics: maximum runtime performance; errors caught at compile time; platform-specific binaries; longer edit-compile-test cycle.

Interpreted Languages

An interpreted language does not require a separate compilation step. An interpreter reads the source code and executes it directly, processing instructions at runtime. You can write a line of code and run it immediately — a workflow that supports rapid experimentation.

Key characteristics: immediate execution without compilation; faster iteration during development; portable across platforms where the interpreter is installed; runtime performance lower than compiled equivalents.

Python: A Hybrid Approach

Python is primarily interpreted, but its execution model is more sophisticated than simple line-by-line interpretation. When you run a Python script, the interpreter first compiles the source code into bytecode — a compact, platform-independent set of instructions. This bytecode is then executed by the Python Virtual Machine (PVM). For performance-critical sections, libraries like NumPy use compiled C under the hood, and tools like PyPy use just-in-time compilation to execute Python code significantly faster than the standard interpreter.

The compiled versus interpreted distinction is rarely the deciding factor when choosing a language. More commonly the decision comes down to available libraries, team skills, deployment environment, and the nature of the problem. Many production systems use both: Python at the application layer, with compiled libraries handling performance-critical components.

1.4 Comparing Python with C++, Java, and C#

Python exists in a landscape of hundreds of programming languages. Four that developers most commonly encounter alongside Python are C++, Java, and C#. Understanding how they compare helps clarify where Python excels and where a different tool might be more appropriate. All four examples below solve the same problem: computing the area of a circle at several radii.

Example 1 — Python

High-level, dynamically typed, interpreted. Minimal ceremony, maximum expressiveness.

```
import math

def circle_area(radius: float) -> float:
    return math.pi * radius ** 2

for r in [1.0, 2.5, 5.0, 10.0]:
    print(f"Radius {r:>5}: area = {circle_area(r):.4f}")
```

Example 2 — C++

Compiled, statically typed, maximum control. More verbose; suited to performance-critical applications.

```
#include <iostream>
#include <cmath>
#include <vector>
using namespace std;

double circle_area(double radius) {
    return M_PI * pow(radius, 2);
}

int main() {
    vector<double> radii = {1.0, 2.5, 5.0, 10.0};
    for (double r : radii)
        cout << "Radius " << r << ": area = " << circle_area(r) << endl;
    return 0;
}
```

Example 3 — Java

Compiled to bytecode, runs on the JVM. Platform-independent; dominant in enterprise and Android.

```
import java.util.Arrays;
import java.util.List;

public class CircleArea {
    public static double circleArea(double radius) {
        return Math.PI * Math.pow(radius, 2);
    }
    public static void main(String[] args) {
        List<Double> radii = Arrays.asList(1.0, 2.5, 5.0, 10.0);
        for (double r : radii)
            System.out.printf("Radius %.1f: area = %.4f%n", r, circleArea(r));
    }
}
```

C# follows a similar structure to Java and is the primary language for Windows application development and Unity game development. Python is the clearest and most concise of the four, with the smallest gap between the idea and the code. C++ is fastest at runtime. Java and C# provide strong compile-time guarantees and mature enterprise ecosystems. The best choice depends on the domain, the performance requirements, and the deployment environment.

1.5 Why Python?

Python has become one of the most widely used programming languages in the world, spanning roles from introductory teaching to primary tool of professional data scientists, infrastructure engineers, and researchers at the world's largest technology organisations. This breadth reflects a set of design choices that make Python genuinely productive across a remarkable range of contexts.

Readable, Low-Ceremony Syntax

Python's syntax is designed to be as close to plain English as practical. It uses indentation to define code blocks, eliminates mandatory semicolons, and avoids much of the structural boilerplate required by languages like Java and C++. Code that is easy to read is easier to review, easier to maintain, and easier to hand off. Compare the same logic in Python and Java:

Example 1 — Introduction

Checking whether a port number is in the valid range — Python vs Java.

```
# Python
def is_valid_port(port: int) -> bool:
    return 1 <= port <= 65535

// Java
public static boolean isValidPort(int port) {
    return port >= 1 && port <= 65535;
}
```

Both are readable, but Python requires less structural knowledge to write and understand. This lower barrier is consistent across the language.

Dynamic Typing and Optional Type Hints

Python is dynamically typed: variables do not need to be declared with a type, and the same variable can hold values of different types at different times. This reduces the amount of code required for exploratory work. Python 3.5 introduced optional type hints, allowing developers to annotate function signatures for documentation and static analysis without losing flexibility:

Example 2 — Variation

The same function written without and with type hints — both are valid Python.

```
# Without type hints -- concise, flexible
def compute_discount(price, rate):
    return price * (1 - rate)

# With type hints -- explicit, better tooling support
def compute_discount(price: float, rate: float) -> float:
    return price * (1 - rate)
```

Community, Ecosystem, and Versatility

Python has one of the largest and most active developer communities of any programming language. The Python Package Index (PyPI) hosts over 500,000 packages covering virtually every domain. The quality of the most widely used packages — NumPy, Pandas, Requests, SQLAlchemy, Django — is exceptional, with large contributor bases, rigorous testing, and stable APIs.

Most languages are strongly associated with one or two primary domains. Python is unusual in being genuinely first-rate across data science, web development, automation, scientific computing, machine learning, and DevOps simultaneously. Skills developed in one domain transfer directly to others, and the same language covers the full workflow from data acquisition through modelling to serving predictions via an API.

1.6 Real-World Uses of Python

Python's presence in production systems across every major industry is one of the clearest measures of its practical value.

Web Development

Django is a full-featured framework that provides an ORM for database interaction, built-in authentication, and URL routing — everything needed to build a production web application. FastAPI is a modern, high-performance framework for REST APIs, notable for generating input validation and interactive documentation automatically from type hints. Instagram, Spotify's backend services, and a wide range of government digital platforms run on Python web frameworks.

Data Science and Analytics

Python has become the standard working language of data science, largely displacing R and proprietary tools like MATLAB over the past decade. The Pandas library provides DataFrame-based data manipulation. NumPy provides high-performance array operations. Matplotlib and Seaborn cover visualisation. Jupyter Notebooks provide an interactive environment where code, results, and narrative prose coexist in a single document — standard in analytical reporting contexts.

Machine Learning and Artificial Intelligence

Python is the dominant language for machine learning research and production deployment. TensorFlow (Google) and PyTorch (Meta) are the two leading deep learning frameworks, both Python-based. Scikit-learn provides a consistent API for classical machine learning algorithms. Hugging Face's Transformers library has made state-of-the-art natural language processing accessible in a few lines of Python. Organisations across banking, retail, healthcare, and logistics use Python-based models in production daily.

Automation and Infrastructure

Python's clear syntax and extensive standard library make it an excellent tool for automation. Infrastructure and operations teams use Python for configuration management (Ansible, SaltStack), cloud resource provisioning via AWS, Google Cloud, and Azure SDKs, and orchestrating complex workflows with tools like Airflow and Prefect. A DevOps engineer might write Python scripts to provision infrastructure, monitor service health, rotate credentials, and send alerts — integrating freely with any API or command-line tool.

Scientific and Research Computing

Across academia and research institutions, Python has replaced proprietary tools as the standard environment for scientific computing. SciPy provides algorithms for optimisation, integration, and signal processing. Astropy supports astronomical data analysis. Biopython supports bioinformatics workflows. Python's open-source, auditable ecosystem supports the reproducibility standards that research communities increasingly require.

1.7 The Python Ecosystem and Tools

Writing Python is not just a matter of the language itself — it involves a surrounding ecosystem of tools, libraries, and environments that shape the development experience. This section covers the components you will encounter most often.

Development Environments

IDLE ships with every Python installation. It provides an interactive shell where expressions can be evaluated immediately and a text editor for writing scripts. It is the fastest way to start — no setup required. Visual Studio Code (VS Code) is the recommended environment for ongoing work. The Python extension adds IntelliSense completion, inline error checking, an integrated debugger, and a built-in terminal. PyCharm is a full-featured Python IDE with more sophisticated refactoring and deeper framework integration, preferred in professional environments with large codebases. Jupyter Notebooks are standard in data science and research, where code, results, and narrative coexist in a single shareable document.

NumPy and Pandas

NumPy is the numerical computing foundation of Python's scientific stack. It provides the ndarray type — a fast, memory-efficient multi-dimensional array — and a comprehensive library of mathematical operations. Because NumPy's core is implemented in compiled C, operations on large arrays execute at speeds that pure Python loops cannot approach.

Example 1 — Introduction

NumPy element-wise operations across two arrays of product data.

```
import numpy as np

prices     = np.array([12.50, 8.99, 22.00, 5.75, 17.30])
quantities = np.array([100, 250, 80, 400, 150])

revenue = prices * quantities
print(f"Revenue per product: {revenue}")
print(f"Total revenue: ${revenue.sum():,.2f}")
```

Pandas builds on NumPy to provide the DataFrame — a two-dimensional labelled data structure equivalent to a database table with powerful built-in capabilities for filtering, grouping, joining, and summarising data.

Example 2 — Variation

Pandas DataFrame to calculate and rank product revenues.

```
import pandas as pd

df = pd.DataFrame({
    'product': ['Widget A', 'Widget B', 'Widget C'],
    'price':   [12.50, 8.99, 22.00],
    'units':   [100, 250, 80]
})

df['revenue'] = df['price'] * df['units']
print(df.sort_values('revenue', ascending=False))
```

Visualisation, Machine Learning, and Web

Matplotlib is Python's foundational plotting library, producing publication-quality charts and graphs. Seaborn builds on Matplotlib with higher-level statistical visualisation and cleaner defaults. For machine learning, TensorFlow (Google) and PyTorch (Meta) provide the leading deep learning frameworks; Scikit-learn covers classical machine learning algorithms with a consistent, well-documented API. For web development, Django provides a full-featured framework with built-in authentication, ORM, and admin interface; FastAPI is the modern choice for high-performance REST APIs.

Example 3 — Real-world application

A Matplotlib bar chart summarising monthly revenue — typical of a business reporting workflow.

```
import matplotlib.pyplot as plt

months  = ["Jan", "Feb", "Mar", "Apr", "May", "Jun"]
revenue = [42000, 38500, 51000, 47800, 55200, 61000]

plt.figure(figsize=(9, 4))
plt.bar(months, revenue, color="steelblue")
plt.title("Monthly Revenue")
plt.ylabel("Revenue ($)")
plt.tight_layout()
```

```
plt.show()
```

Version Control: Git

Version control is not Python-specific, but no programming workflow is complete without it. Git is the universal standard for tracking changes to code over time, enabling multiple people to work on the same project simultaneously and maintaining a complete history with the ability to revert to any prior state. GitHub, GitLab, and Bitbucket host Git repositories. From the first script you write, committing work to version control is one of the most valuable professional habits you can build.

Chapter 1 Summary

- Programming is writing precise instructions for a computer to follow. Every program reads data, transforms it, and produces a result.
- Python supports three major paradigms: procedural (step-by-step functions), object-oriented (data and behaviour bundled in objects), and functional (data transformed through pure functions without side effects).
- Compiled languages translate source code to machine code before running; interpreted languages execute at runtime. Python uses bytecode as an intermediate step, balancing portability with reasonable performance.
- Compared with C++, Java, and C#, Python's syntax is the most concise and readable. Each language has domains where it excels: C++ for performance, Java/C# for managed enterprise ecosystems, Python for data, automation, and rapid development.
- Python's combination of readable syntax, dynamic typing, a 500,000-package ecosystem, and genuine versatility across domains makes it one of the most productive languages for a wide range of real-world tasks.
- Python is used professionally in web development (Django, FastAPI), data science (Pandas, NumPy), machine learning (TensorFlow, PyTorch), automation and DevOps, and scientific research.
- Core tools: IDLE for quick starts, VS Code for day-to-day work, PyCharm for large projects, Jupyter Notebooks for data work. Git for version control from day one.

Guided Exercises

Complete the following exercises in VS Code, IDLE, or your preferred Python environment. Each exercise targets a specific concept from this chapter.

1. Write a Python function that accepts a temperature in Celsius and returns the equivalent in Fahrenheit. Call the function for 0°C, 20°C, 37°C, and 100°C and print each result with one decimal place.

2. Write a program that takes a list of five product prices and prints the total, the minimum price, and the maximum price using Python's built-in functions. Then add a 10% discount to each price and print the discounted list.

3. Use an f-string to print a formatted table of three currencies and their exchange rates against AUD, with the currency name left-aligned in a column of 8 characters and the rate right-aligned with two decimal places.

4. Write a function with a complete docstring that accepts a list of numbers and returns the average, handling the case where the list is empty. Call the function with at least three different lists and print each result.

5. Rewrite the temperature analysis code from Section 1.1 so that: (a) it uses a named constant for the threshold, (b) the analysis is encapsulated in a function, and (c) the output includes the percentage of days over threshold as well as the count.

Challenge Tasks

The following tasks are open-ended. No hints are provided. They require you to design a solution, not just apply a pattern.

1. Design and implement a simple command-line currency converter. It should: store exchange rates in a dictionary; accept an amount and a target currency code from the user; validate both inputs with clear error messages; and display the converted result formatted to two decimal places. Demonstrate the converter with at least four currency pairs.

2. Demonstrate all three programming paradigms from Section 1.2 by solving the same problem three different ways. The problem: given a list of product prices, compute the total, the average, and the count of items priced above the average. Write one procedural version, one OOP version (with a class), and one functional version using filter(), map(), or reduce().

3. Using the NumPy and Pandas concepts from Section 1.7, build a small data analysis script that: loads at least six rows of realistic data (you may define it directly in the script as a dictionary or list); computes summary statistics (mean, max, min, and total); filters the data to a subset based on a condition; and prints a formatted report. Choose a real-world domain: sales figures, sensor readings, sports statistics, or similar.

CHAPTER 2 **Writing Your First Python Script**

Think, Code, Create – A Python Programming Journey

With the conceptual groundwork of Chapter 1 established, it is time to write and execute actual Python code. This chapter covers the practical foundations that every working Python script depends on: how Python files are structured, how to produce output and receive input, how indentation defines the shape of a program, and how to work effectively in a Python development environment.

Section 2.8 introduces running Python from the command line and terminal — a skill that moves you beyond relying on any single editor and is standard practice in professional and automated workflows. By the end of this chapter you will be able to write, save, and execute Python scripts with clarity and intention, and recover confidently from the errors that every programmer encounters along the way.

2.1 Understanding the Python File Structure

A Python source file is a plain text file with a .py extension. Any text editor can create one. What matters is that the content follows Python's syntax rules and is passed to the Python interpreter for execution. Understanding the conventional structure of a Python file helps you write code that is easy to read, navigate, and maintain as projects grow.

The Basic Python File

At its simplest, a Python script is a sequence of statements executed from top to bottom. The interpreter reads the file, processes each statement in order, and halts when it reaches the end or encounters an unhandled error.

Example 1 — Introduction

A minimal script that reads a sensor value and reports whether it is within an acceptable range.

```
# sensor_check.py
# Reads a sensor value and reports whether it is within the acceptable range.

LOWER_BOUND = 0.0
UPPER_BOUND = 100.0

raw   = input("Enter sensor reading: ")
value = float(raw)

if LOWER_BOUND <= value <= UPPER_BOUND:
    print(f"Reading {value} is within range.")
else:
    print(f"Reading {value} is OUT OF RANGE. Check sensor.")
```

Constants are defined at the top, input is collected and converted, a decision is made, and output is produced. This top-to-bottom flow is the foundation of every Python program.

Conventional File Structure

As scripts grow, a consistent internal structure keeps them navigable. The widely adopted convention is: imports first (all dependencies visible immediately); constants and configuration next (uppercase names signal their purpose); function and class definitions (the reusable logic, defined before the code that calls them); and finally the main execution block.

Example 2 — Variation

A report generator structured with imports, constants, functions, and a guarded main block.

```
# report_generator.py
import csv
from datetime import date

REPORT_PATH = "monthly_report.txt"
DATA_FILE   = "transactions.csv"

def load_transactions(filepath):
    with open(filepath, newline="") as f:
        return list(csv.DictReader(f))

def summarise(transactions):
    return sum(float(t["amount"]) for t in transactions)

if __name__ == "__main__":
    data  = load_transactions(DATA_FILE)
    total = summarise(data)
    print(f"Report for {date.today()}: total = ${total:,.2f}")
```

The if __name__ == '__main__': guard ensures the main block only runs when the script is invoked directly, not when it is imported as a module by another file — a distinction that becomes important as projects grow.

Modules and Packages

When a project grows beyond a single file, code is split across multiple .py files called modules. A package is a directory containing an __init__.py file and a collection of related modules. Imports draw on the standard library, third-party packages, or local modules:

Example 3 — Real-world application

Import patterns encountered in a real project: standard library, third-party, and local modules.

```
import os
import json
from pathlib import Path          # standard library
from requests import get          # third-party (pip install requests)
from utils import format_currency  # local module in the same project
```

2.2 print() Statements and Output

The print() function is the primary mechanism for producing output in a Python script. It sends text to standard output — the terminal where the script is running — and is indispensable both for presenting results to a user and for inspecting program behaviour during development.

Basic Usage and the sep Parameter

print() accepts one or more arguments separated by commas and writes them to the terminal followed by a newline. Multiple arguments are separated by a single space by default; the sep parameter overrides this:

Example 1 — Introduction

Using sep to format a date string and a structured log line.

```
print("2024", "07", "15", sep="-")          # 2024-07-15
print("INFO", "Service started", sep=" | ") # INFO | Service started


server = "db-prod-01"
port   = 5432
print("Connecting to", server, "on port", port)
```

The end Parameter

By default print() appends a newline after the output. The end parameter overrides this, which is useful when building output progressively across multiple calls:

```
import time

steps = ["Connecting", "Authenticating", "Fetching data", "Done"]
for step in steps:
    print(step, end=" ... ")
    time.sleep(0.3)
print()  # final newline
```

f-strings

f-strings (formatted string literals, available from Python 3.6) are the recommended way to embed values and expressions into output strings. Prefix the string with f and place any Python expression inside curly braces. Common format specifiers: :, for thousands separator; :.2f for two decimal places; :>10 for right-alignment in a field of 10 characters.

Example 2 — Variation

f-string formatting applied to a data processing summary and an aligned product table.

```
filename   = "export_2024_Q3.csv"
row_count  = 14_872
elapsed_ms = 312.7
print(f"Processed {filename}: {row_count:,} rows in {elapsed_ms:.1f}ms")
# Processed export_2024_Q3.csv: 14,872 rows in 312.7ms


items = [("CPU", 4), ("RAM (GB)", 16), ("Disk (TB)", 2)]
for label, value in items:
```

```
    print(f"  {label:<12} {value:>4}")
```

Printing to Standard Error

Diagnostic messages and errors should be directed to standard error (stderr) rather than standard output, so they can be distinguished from program output when scripts are composed in pipelines. Pass file=sys.stderr to print():

Example 3 — Real-world application

A file-processing function that directs success output to stdout and errors to stderr.

```
import sys

def process_file(path):
    try:
        with open(path) as f:
            data = f.read()
        print(f"Read {len(data)} bytes from {path}")
    except FileNotFoundError:
        print(f"ERROR: File not found: {path}", file=sys.stderr)
```

2.3 Basic Syntax and Indentation Rules

Python's syntax is designed for readability. Compared to languages that use curly braces and semicolons to delimit statements and blocks, Python uses whitespace and a small set of punctuation conventions. These rules are strict — the interpreter enforces them precisely — but once internalised they produce code that reads almost like structured prose.

Statements and Line Continuation

Each logical statement in Python occupies its own line. When a statement is too long to fit comfortably, continue it inside open brackets — the preferred approach — or use a trailing backslash:

```
# Continuation inside brackets -- preferred
total = (
    base_salary
    + overtime_pay
    + performance_bonus
    - tax_withheld
)
```

Indentation Defines Structure

Python uses indentation to define code blocks. Every statement that belongs to a block — the body of a function, the branch of a conditional, the body of a loop — must be indented one level relative to the statement that introduces the block. The standard is four spaces per indentation level. Incorrect indentation causes either an IndentationError or, worse, a logic error where the wrong statements belong to the wrong block.

Example 1 — Introduction

Correct indentation across a function containing a loop and a nested conditional.

```
def check_inventory(items, threshold):
    low_stock = []
    for item in items:
        if item["quantity"] < threshold:
            low_stock.append(item["name"])
    return low_stock

stock = [
    {"name": "Widget A", "quantity": 42},
    {"name": "Widget B", "quantity": 3},
    {"name": "Widget C", "quantity": 17},
]
needs_reorder = check_inventory(stock, threshold=10)
print("Reorder required:", needs_reorder)
```

Colons, Comments, and Docstrings

Any statement that introduces a block ends with a colon: function definitions, class definitions, conditionals, loops, and exception handlers. Forgetting the colon is one of the most common syntax errors. Single-line comments begin with # and are ignored by the interpreter. Docstrings — triple-quoted strings as the first statement of a function or class — provide structured, machine-readable documentation:

Example 2 — Variation

A function demonstrating correct colon use, an inline comment, and a complete docstring.

```
def calculate_tax(income: float, rate: float) -> float:
    """
    Calculate tax payable on a given income at a flat rate.

    Args:
        income: Gross income in dollars.
        rate:   Tax rate as a decimal (e.g. 0.325 for 32.5%).
    Returns:
        Tax payable as a float.
    """
    return income * rate

RATE = 0.325  # Current marginal rate for this bracket
```

Naming Conventions

Python follows PEP 8, the official style guide. Key naming rules: variables and functions use lowercase_with_underscores (e.g. user_id, process_batch); constants use UPPERCASE_WITH_UNDERSCORES (e.g. MAX_RETRIES, API_BASE_URL); classes use PascalCase (e.g. DatabaseConnection, InvoiceProcessor). Choosing names that describe purpose rather than structure is the single most effective thing you can do to improve readability.

Example 3 — Real-world application

Comparing opaque names against descriptive PEP 8-compliant names for the same variables.

```
# Hard to interpret
x  = 86400
y  = t / x

# Self-documenting
SECONDS_PER_DAY = 86400
elapsed_days    = total_seconds / SECONDS_PER_DAY
```

2.4 Writing and Running Python Scripts

Python scripts can be written and executed from several environments. The right choice depends on the complexity of the work, the project structure, and personal preference. This section covers the three environments you are most likely to use: IDLE, Visual Studio Code, and the terminal.

IDLE

IDLE ships with every Python installation and requires no configuration. It provides two components: an interactive shell where expressions can be evaluated immediately, and a text editor for writing scripts. The interactive shell is useful for quick experiments — testing a single expression or checking the behaviour of a function:

Example 1 — Introduction

Using the IDLE interactive shell to explore the datetime module before writing a script.

```
>>> from datetime import date
>>> date.today()
datetime.date(2024, 7, 15)
>>> date.today().strftime("%d %B %Y")
'15 July 2024'
```

To write a longer script, open File → New File, type the code, save with a .py extension, and press F5 (or Run → Run Module) to execute it. Output appears in the shell window.

Visual Studio Code

VS Code is the recommended environment for ongoing Python work. Install the Python extension (published by Microsoft) and VS Code gains IntelliSense code completion, inline type checking, an integrated debugger, and a built-in terminal. To run the current file, open the integrated terminal (Ctrl+`) and type python filename.py, or use the Run button in the top-right corner. Set breakpoints by clicking the gutter beside a line number, then run in debug mode to pause execution and inspect variables at any point.

Virtual Environments

Virtual environments isolate a project's dependencies so that installing a package for one project does not affect another. VS Code detects a .venv directory automatically and offers to use it as the project interpreter:

Example 2 — Variation

Creating, activating, and using a virtual environment for an isolated project.

```
# Create the environment
python -m venv .venv

# Activate -- macOS / Linux
source .venv/bin/activate

# Activate -- Windows
.venv\Scripts\activate

# Install packages into the environment only
pip install requests pandas
```

2.5 Escape Characters and Comments

Two features of Python's string handling and code notation deserve dedicated attention: escape characters, which allow special characters to be embedded in strings, and comments, which allow explanatory text to coexist with executable code. Both are used constantly in practice.

Escape Characters

An escape sequence is a backslash followed by a character that together represent a single special character. The most frequently used are: \n (newline), \t (horizontal tab), \\ (literal backslash), and \r (carriage return). Raw strings prefixed with r treat every backslash literally — the preferred approach for Windows file paths and regular expressions.

Example 1 — Introduction

Escape sequences used to format a structured log output with tabs and newlines.

```
header = "Timestamp\tEvent\tSeverity"
row    = "2024-07-15 09:14:02\tDisk full\tCRITICAL"
print(header)
print(row)

# Raw string for a Windows path -- backslashes treated literally
path = r"C:\Users\analyst\data\report.csv"
print(path)
```

Example 2 — Variation

Triple-quoted strings for a multi-line SQL query — no escape characters needed.

```
sql_query = """
    SELECT
        customer_id,
        SUM(amount) AS total_spend
    FROM orders
    WHERE status = 'completed'
    GROUP BY customer_id
    ORDER BY total_spend DESC
    LIMIT 20
"""
```

```
print(sql_query)
```

Comments in Practice

Comments serve two purposes: explaining why code does something (the most valuable kind), and explaining what it does (useful when the logic is non-obvious). Comments that merely restate the code add noise without value. For functions, classes, and modules, docstrings are the correct documentation vehicle — they travel with the code and are surfaced by help(), IDE tooltips, and documentation generators.

Example 3 — Real-world application

Useful inline comments versus a restatement comment, and the docstring as primary documentation.

```
# USEFUL -- explains intent not apparent from the code
RETRY_LIMIT = 3  # AWS recommends up to 3 retries for transient errors

# NOT USEFUL -- restates the obvious
x = x + 1  # increment x by 1

def parse_iso_date(date_str: str):
    """
    Parse an ISO 8601 date string (YYYY-MM-DD) and return a date object.
    Raises ValueError if the string does not match the expected format.
    """
    from datetime import datetime
    return datetime.strptime(date_str, "%Y-%m-%d").date()
```

2.6 Errors and Debugging

Errors are a normal part of writing software. Even experienced developers spend a significant portion of their time reading error messages and tracing the source of unexpected behaviour. Developing a systematic approach to debugging early is one of the highest-leverage habits you can build.

Syntax Errors

A syntax error means the interpreter cannot parse the code — the text does not conform to Python's grammar. Python detects syntax errors before executing anything and reports the file, line number, and a caret pointing to the problem. Common causes: missing colons at the end of if/def/for lines, unmatched brackets, and incorrect indentation.

Example 1 — Introduction

A missing colon triggers a SyntaxError before the script runs.

```
# Missing colon -- Python refuses to execute
if temperature > 100
    print("Overheating")

#   File "monitor.py", line 2
#       if temperature > 100
#                          ^
#   SyntaxError: expected ':'
```

Read the caret position carefully — it often points to the character immediately after the actual mistake. Here the missing colon belongs at the end of the if line.

Runtime Errors (Exceptions)

A runtime error occurs during execution when an operation cannot be completed. Python raises an exception and, if it is not caught, halts with a traceback showing the full call stack. Read tracebacks from the bottom up: the last line names the exception type and the specific value that caused it; the lines above show exactly where in the code it occurred.

Example 2 — Variation

A KeyError traceback and the two idiomatic ways to handle it safely.

```
rates    = {"USD": 1.52, "EUR": 1.63, "JPY": 0.0099}
currency = "GBP"

# This raises: KeyError: 'GBP'
converted = 100 * rates[currency]

# Safe alternatives:
converted = 100 * rates.get(currency, 0)    # returns 0 if key missing

try:
    converted = 100 * rates[currency]
except KeyError:
    print(f"Unknown currency: {currency}")
```

Logic Errors

A logic error produces no exception — the program runs to completion but produces the wrong result. These are the most demanding errors to diagnose. The only way to detect them is to compare actual output against independently known correct output. Test functions with inputs whose correct result you can verify by hand, then work backwards from the discrepancy.

```
def compound_interest(principal, rate, years):
    # Bug: should be (1 + rate), not (1 - rate)
    return principal * (1 - rate) ** years

print(compound_interest(10000, 0.05, 10))
# Prints 5987.37 instead of the correct 16288.95
```

Debugging Techniques

Read the full traceback before changing anything — the exception type, message, and line number together usually identify the problem directly. For logic errors, add targeted print() statements to inspect variable values at the suspected location. Prefix them with DEBUG: so they are easy to find and remove later.

Example 3 — Real-world application

Using targeted debug prints to trace a discount calculation and the VS Code debugger as an alternative.

```
def calculate_discount(price, code):
    discounts = {"SAVE10": 0.10, "SAVE20": 0.20}
    print(f"DEBUG: price={price}, code={code}")   # temporary
    rate = discounts[code]
    print(f"DEBUG: rate={rate}")                  # temporary
    return price * (1 - rate)
```

Both IDLE and VS Code provide interactive debuggers. Set a breakpoint on any line, run in debug mode, and execution pauses so you can inspect all variable values, step line by line, and step into function calls. This is more efficient than inserting print() statements for complex problems. When stuck, articulating the problem out loud — even to yourself — forces a systematic description that often reveals the flaw before any tools are needed.

2.7 Best Practices

The difference between code that works and code that is good is not just correctness — it is readability, consistency, and the ease with which it can be understood, tested, and modified. These qualities are built through deliberate habits applied from the start.

Use Descriptive, Purposeful Names

Variable and function names should communicate intent without requiring the reader to trace the logic. Avoid names like data, temp, val, and result unless the scope is so narrow that the meaning is unambiguous.

Example 1 — Introduction

Opaque variable names versus self-documenting names for the same calculation.

```
# Hard to interpret
x = 86400
y = t / x

# Self-documenting
SECONDS_PER_DAY = 86400
elapsed_days    = total_seconds / SECONDS_PER_DAY
```

Avoid Hardcoded Values

Values that control behaviour — timeouts, limits, file paths, rate constants — should be named constants or configuration parameters. When a value needs to change, a named constant requires one edit in one place; a literal requires finding every occurrence.

Example 2 — Variation

A retry limit hardcoded as a literal versus extracted as a named constant.

```
# Avoid: magic number with no explanation
if retries > 3:
    raise TimeoutError

# Preferred: named constant with context
```

```
MAX_RETRIES = 3
if retries > MAX_RETRIES:
    raise TimeoutError(f"Failed after {MAX_RETRIES} attempts")
```

Keep Functions Focused and Validate Input

A function should do one thing and do it completely. If you find yourself naming a function process_and_save_and_notify, split it. Programs that accept external input should never assume that input is valid — validate before processing and provide clear error messages on failure.

Example 3 — Real-world application

A focused validation function with a docstring, clear error messages, and a constrained return contract.

```
def parse_percentage(raw: str) -> float:
    """
    Convert a string like "85" or "85.5" to a float between 0 and 100.
    Raises ValueError with a descriptive message on invalid input.
    """
    try:
        value = float(raw)
    except ValueError:
        raise ValueError(f"Expected a number, got: {raw!r}")
    if not 0.0 <= value <= 100.0:
        raise ValueError(f"Percentage must be 0-100, got: {value}")
    return value
```

Don't Repeat Yourself

If the same logic appears in two or more places, extract it into a function. Duplicated code creates duplicated maintenance — every bug fix must be applied in multiple locations, and the risk of inconsistency compounds with every edit.

Write for the reader: format code as if the next person seeing it has the skills but not the context. Follow PEP 8 conventions — four spaces of indentation, spaces around operators, blank lines between logical sections, lines no longer than 88 characters. Run your code after every meaningful addition rather than only when the entire program is complete. When something breaks, the cause is confined to the lines you just added.

2.8 Running Python from the Command Line

The terminal is the most direct and portable way to run Python scripts. It is the standard execution environment in automated systems, servers, CI/CD pipelines, and scheduled tasks — contexts where no graphical editor is available and scripts are invoked programmatically. Fluency with the command line is a core professional skill.

Running a Script and Passing Arguments

Navigate to the directory containing your script and invoke the interpreter with the filename. Arguments are passed after the script name and accessed inside the script via sys.argv, a list of strings where sys.argv[0] is the script name and subsequent elements are the arguments provided:

Example 1 — Introduction

A command-line currency converter that validates sys.argv input and exits with an error code on failure.

```
# convert.py
import sys

if len(sys.argv) != 3:
    print("Usage: python convert.py <amount> <currency>", file=sys.stderr)
    sys.exit(1)

RATES = {"AUD": 1.52, "EUR": 1.07, "GBP": 0.81, "JPY": 157.4}

amount   = float(sys.argv[1])
currency = sys.argv[2].upper()

if currency not in RATES:
    print(f"Unknown currency: {currency}", file=sys.stderr)
    sys.exit(1)

print(f"USD {amount:.2f} = {currency} {amount * RATES[currency]:.2f}")

# Invoked as:  python convert.py 250 AUD
# Output:      USD 250.00 = AUD 380.00
```

Interactive Mode and Exit Codes

Running python with no arguments starts the interactive interpreter — the same shell IDLE provides, but in the terminal. The >>> prompt indicates Python is ready for input. When a script finishes normally it exits with code 0 (success); a sys.exit(1) call or unhandled exception produces a non-zero code (failure). Exit codes matter when scripts are called by orchestration tools or composed in shell pipelines.

The Shebang Line

On Linux and macOS, a script can be made directly executable by adding a shebang as the first line and setting the executable permission. The #!/usr/bin/env python3 form uses env to locate python3 in the system's PATH, making the script portable across machines with Python installed in different locations:

Example 2 — Variation

A self-executable script with a shebang line — standard for server scripts, cron jobs, and containers.

```
#!/usr/bin/env python3
# daily_report.py

import datetime
print(f"Daily report: {datetime.date.today()}")

# Make executable and run directly:
# chmod +x daily_report.py
# ./daily_report.py
# Daily report: 2024-07-15
```

Exit Codes in Automated Workflows

Example 3 — Real-world application

A config loader that signals failure to the calling process with a non-zero exit code.

```
#!/usr/bin/env python3
import sys, json

def main():
    config_path = "config.json"
    try:
        with open(config_path) as f:
            config = json.load(f)
    except FileNotFoundError:
        print(f"Config file not found: {config_path}", file=sys.stderr)
        sys.exit(1)
    except json.JSONDecodeError as e:
        print(f"Invalid JSON in {config_path}: {e}", file=sys.stderr)
        sys.exit(1)

    print(f"Environment: {config.get('environment', 'unknown')}")

if __name__ == '__main__':
    main()
```

Chapter 2 Summary

- A Python source file is a plain text file with a .py extension. The conventional structure is: imports, constants, function/class definitions, then a guarded main block using if __name__ == '__main__':.
- print() sends output to the terminal. Key parameters: sep controls the separator between arguments (default space); end controls what follows the last argument (default newline). f-strings embed expressions directly into strings and support format specifiers for alignment, decimal places, and thousands separators.
- Direct error and diagnostic messages to sys.stderr to keep them separate from program output — critical when scripts are composed in pipelines or called by orchestration tools.
- Python uses indentation to define code blocks — four spaces per level is the standard. Any statement introducing a block ends with a colon. Inconsistent indentation causes IndentationError.
- Escape sequences embed special characters in strings: \n (newline), \t (tab), \\ (backslash). Raw strings (prefix r) treat backslashes as literals. Triple-quoted strings span multiple lines without escape characters.
- Three error types: syntax errors (caught before execution, grammar violation), runtime errors/exceptions (raised during execution, traceback shows cause and location), and logic errors (program runs but produces wrong output — found through testing and targeted print() statements or a debugger).
- Best practices: descriptive names, named constants instead of magic numbers, focused single-purpose functions, validated input, no repeated logic, and incremental testing after each meaningful addition.

- Scripts are invoked from the terminal with python filename.py. Command-line arguments are accessed via sys.argv. A non-zero sys.exit() signals failure to the calling process. The shebang line (#!/usr/bin/env python3) makes a script self-executable on Linux and macOS.

Guided Exercises

Complete the following exercises in VS Code, IDLE, or your preferred Python environment. Each exercise targets a specific concept from this chapter.

1. Write a script structured with imports, constants, two functions, and a main block. The script should define a constant TAX_RATE, a function compute_tax(price) that returns the tax on a given price, and a function print_receipt(item, price) that prints a formatted receipt showing the item, price excluding tax, tax amount, and total. Call print_receipt for at least three items in the main block.

2. Use f-string format specifiers to print a two-column table of at least five server names and their response times in milliseconds, with the server name left-aligned in a 20-character field and the response time right-aligned in an 8-character field with one decimal place.

3. Write a function parse_percentage(raw) that accepts a string, validates it is a number between 0 and 100, and returns a float. Raise a ValueError with a descriptive message if validation fails. Include a complete docstring. Test it with at least two valid and two invalid inputs, printing the result or the error message for each.

4. Demonstrate three escape sequences and a raw string in a single script: print a two-column tab-aligned table using \t; print a multi-line address block using \n within a single string; and print a Windows-style file path using a raw string.

5. Write a script that accepts a filename as a command-line argument using sys.argv. If the argument is missing, print a usage message to stderr and exit with code 1. If the file does not exist, print an error to stderr and exit with code 1. If the file exists, print the number of lines, words, and characters it contains.

Challenge Tasks

The following tasks are open-ended. No hints are provided. They require you to design a solution, not just apply a pattern.

1. Build a command-line environment checker: a script that a new team member can run immediately to verify their Python setup is correct. It should check the Python version (>= 3.9

required), confirm that at least three specified libraries are importable (e.g. requests, pandas, numpy), verify that a config.json file exists in the current directory and contains required keys, and print a clear PASS or FAIL summary for each check with a final overall result. Use appropriate exit codes.

2. Write a script that reads its own source file (hint: __file__ gives the script's path), then reports: total line count, number of comment lines (starting with #), number of blank lines, number of def lines (function definitions), and number of code lines (everything else). Print a formatted summary. Verify that the counts add up correctly.

3. Design and build a structured logging utility module (logger.py) with three functions: log_info(message), log_warning(message), and log_error(message). Each function should write a timestamped, labelled line to the appropriate stream (info and warning to stdout, error to stderr) in the format: [YYYY-MM-DD HH:MM:SS] LEVEL message. Write a second script (app.py) that imports and uses all three functions, demonstrating correct separation of concerns between the two files.

CHAPTER 3 **Variables**

Think, Code, Create – A Python Programming Journey

> Variables are the primary mechanism by which a Python program stores, tracks, and transforms information. Without them, every value would be a hard-coded literal — fixed, unreusable, and impossible to respond to external input. With them, a program can remember a running total, hold a user's response, compute an intermediate result, and carry that result forward into the next operation.
>
> This chapter covers everything you need to work with variables confidently: how to create and name them, the four fundamental data types they carry, how to convert between those types, how Python's dynamic typing system works and what its implications are, and how variables are combined into the expressions that do the computational work of a program. Real-world examples throughout show variables being used in the kinds of contexts you will actually encounter in practice.

3.1 Introduction to Variables

A variable is a named reference to a value stored in memory. When you write temperature = 22.5, Python allocates memory to hold the float 22.5 and binds the name temperature to that location. From that point forward in the program, whenever you write temperature, Python retrieves the value at that memory location. When you assign a new value to the same name, the binding is updated — the old value is released and the name now refers to the new one.

Python is dynamically typed, which means variables do not need to be declared with a type before use. The type is inferred from the value assigned. This is fundamentally different from statically typed languages like Java or C++, where every variable's type must be stated explicitly. In Python, the same name can hold an integer on one line and a string on the next.

Creating and Updating Variables

Example 1 — Introduction

Variables tracking the state of a server monitoring script.

```
# Server monitoring state
hostname        = "api-prod-01"
response_time   = 312.4       # milliseconds
is_healthy      = True
request_count   = 0

# Variables can be updated at any point
request_count   = request_count + 1
response_time   = 289.7

print(f"{hostname}: {response_time}ms, requests: {request_count}")
```

Notice that response_time started as 312.4 and was later updated to 289.7. The name stays the same; the value it points to changes. This mutability is what makes variables useful for tracking state that evolves as a program runs.

The type() Function and isinstance()

Python provides two built-in tools for inspecting the type of a value. type() returns the exact class of an object. isinstance() checks whether an object is an instance of a given class or any of its subclasses. In practice, isinstance() is preferred for type-checking in code because it handles inheritance correctly and is more robust:

Example 2 — Variation

type() for inspection during development; isinstance() for type-checking in production code.

```
price    = 49.95
quantity = 3
label    = "Widget A"

# type() -- useful for debugging and inspection
print(type(price))    # <class 'float'>
print(type(quantity)) # <class 'int'>
print(type(label))    # <class 'str'>

# isinstance() -- preferred for type-checking in code
def apply_discount(price, rate):
    if not isinstance(price, (int, float)):
        raise TypeError(f"price must be numeric, got {type(price).__name__}")
    return price * (1 - rate)

print(apply_discount(49.95, 0.10))  # 44.955
```

Constants

Python has no built-in constant enforcement, but the convention of writing constants in UPPER_CASE signals to any reader that a value is not intended to change. By collecting constants at the top of a file, you create a single point of control for every configurable value in the program:

Example 3 — Real-world application

A shipping rate calculator with all configurable values defined as named constants.

```
STANDARD_RATE_PER_KG = 4.50
EXPRESS_RATE_PER_KG  = 9.80
FREE_SHIPPING_THRESHOLD = 150.00
GST_RATE             = 0.10

def calculate_shipping(weight_kg, order_total, express=False):
    if order_total >= FREE_SHIPPING_THRESHOLD:
        return 0.0
    rate = EXPRESS_RATE_PER_KG if express else STANDARD_RATE_PER_KG
    base = weight_kg * rate
    return round(base * (1 + GST_RATE), 2)

print(calculate_shipping(2.5, 80.00))          # 12.38
print(calculate_shipping(2.5, 80.00, express=True))  # 26.95
```

```
print(calculate_shipping(2.5, 200.00))          # 0.0 -- free
```

3.2 Naming Conventions

A name is the first piece of documentation any reader encounters. A well-chosen name makes the purpose of a variable immediately clear without requiring the reader to trace the logic. A poorly chosen name forces the reader to hold additional context in their head just to understand what the code is doing. Python's official style guide (PEP 8) defines the naming conventions followed throughout this book and expected in professional Python codebases.

Variables and Functions

Use lowercase_with_underscores (snake_case). Names should describe what the variable holds, not how it is stored or what type it is. Avoid single-letter names except in the tightest of local scopes (a loop counter over two lines). Avoid abbreviations that are not universally understood.

Example 1 — Introduction

Opaque names versus descriptive snake_case names for the same data.

```
# Hard to interpret without tracing the logic
ts  = 1_720_051_200
uct = 0.032
ov  = 1.15

# Self-documenting
unix_timestamp       = 1_720_051_200
underwriting_cost    = 0.032
overhead_multiplier  = 1.15
```

Constants, Classes, and Private Members

Constants use UPPER_CASE_WITH_UNDERSCORES, making them immediately distinguishable from variables that change. Class names use PascalCase (each word capitalised, no underscores). Private members — attributes and methods intended for internal use only — are prefixed with a single underscore, signalling to other developers that they should not be accessed directly from outside the class. A double underscore prefix triggers Python's name-mangling mechanism, making the attribute harder (though not impossible) to access from subclasses or external code:

Example 2 — Variation

Naming conventions demonstrated across constants, a class, and private attributes in one example.

```
MAX_CONNECTION_RETRIES = 3          # UPPER_CASE constant
DEFAULT_TIMEOUT_SECONDS = 30

class DatabaseConnection:           # PascalCase class name
    def __init__(self, host, port):
        self.host   = host          # public attribute
        self.port   = port          # public attribute
        self._pool  = []            # _single: internal use only
        self.__secret_key = None    # __double: name-mangled
```

```
    def connect(self):                # lowercase_with_underscores method
        pass

    def _validate_host(self):         # _single: internal helper
        pass
```

Avoiding Ambiguity

Names should be specific enough that their meaning is unambiguous without reading surrounding code. The names data, info, result, temp, and value are almost always too vague for anything beyond the smallest scope. The single characters l (lowercase L), O (uppercase o), and I (uppercase i) should never be used as variable names because they are visually indistinguishable from 1 and 0 in most fonts.

Example 3 — Real-world application

Refactoring ambiguous names in a payroll calculation function to descriptive ones.

```
# Before: ambiguous names
def calc(e, h, r, ot):
    return (h * r) + (ot * r * 1.5)

# After: self-documenting
def calculate_weekly_pay(employee_id, hours_worked, hourly_rate,
overtime_hours):
    regular_pay  = hours_worked * hourly_rate
    overtime_pay = overtime_hours * hourly_rate * 1.5
    return regular_pay + overtime_pay

gross = calculate_weekly_pay("E1042", 38, 32.50, 3)
print(f"Gross pay: ${gross:.2f}")  # Gross pay: $1380.75
```

3.3 Data Types: int, float, str, bool

Every value in Python has a type, and the type determines what operations are valid on that value and how it behaves in expressions. The four types covered in this section — integers, floats, strings, and booleans — are the building blocks that appear in virtually every Python program.

Integers (int)

An integer is a whole number with no fractional component: positive, negative, or zero. Python integers have arbitrary precision — there is no fixed upper limit, unlike many other languages. Integer literals can include underscores as thousands separators for readability without affecting the value. Integer division with // returns a floored integer; regular division with / always returns a float even when the result is whole.

Example 1 — Introduction

Integer arithmetic in a stock inventory context.

```
units_in_stock    = 1_482
```

```python
units_on_order   = 500
units_reserved   = 73

available         = units_in_stock - units_reserved
total_after_order = available + units_on_order
cartons           = available // 24   # integer floor division
remainder         = available % 24    # units left over

print(f"Available: {available}")
print(f"Full cartons: {cartons}, loose units: {remainder}")
```

Floating-Point Numbers (float)

A float is a number with a fractional component. Python floats follow the IEEE 754 double-precision standard, which means they can represent an enormous range of values but cannot represent every decimal fraction exactly. The value 0.1 + 0.2 in Python produces 0.30000000000000004, not 0.3, because neither 0.1 nor 0.2 has an exact binary representation. For financial calculations where exactness matters, use the decimal module instead.

Example 2 — Variation

Float arithmetic for a fuel cost calculation, and the Decimal alternative for financial precision.

```python
# Standard float -- sufficient for most scientific/engineering use
distance_km      = 847.3
fuel_efficiency  = 12.4    # litres per 100 km
price_per_litre  = 2.089

fuel_used  = (distance_km / 100) * fuel_efficiency
total_cost = fuel_used * price_per_litre
print(f"Fuel: {fuel_used:.1f}L, Cost: ${total_cost:.2f}")

# Decimal -- for financial calculations requiring exact representation
from decimal import Decimal
price    = Decimal("19.99")
quantity = 3
total    = price * quantity
print(f"Invoice total: ${total}")  # 59.97 (exact)
```

Strings (str)

A string is an ordered sequence of characters. Strings are immutable — once created, the characters cannot be changed in place, only new strings can be produced from them. String literals can be delimited by single quotes, double quotes, or triple quotes (for multi-line strings). Strings support concatenation with +, repetition with *, indexing, and slicing, as well as a rich library of built-in methods.

Example 3 — Real-world application

String construction and common methods applied to log record processing.

```python
raw_log = "  2024-07-15T09:14:02 | ERROR | disk quota exceeded  "

cleaned   = raw_log.strip()            # remove surrounding whitespace
parts     = cleaned.split(" | ")       # split on delimiter
```

```
timestamp = parts[0]
level     = parts[1]
message   = parts[2].lower()

print(f"[{level}] {timestamp}: {message}")
# [ERROR] 2024-07-15T09:14:02: disk quota exceeded

alert = f"ALERT raised at {timestamp}" if level == "ERROR" else "No alert"
print(alert)
```

Booleans (bool)

A boolean holds exactly one of two values: True or False. Booleans are the result of comparison and logical operations and drive the decision-making logic of every conditional statement and loop. In Python, bool is a subclass of int — True has the integer value 1 and False has the value 0, which means they can participate in arithmetic (though this is rarely a good idea in practice). Any value can be evaluated in a boolean context: zero, empty containers, and None are all falsy; everything else is truthy.

```
is_authenticated = True
is_admin         = False
session_active   = True

# Booleans from comparisons
balance          = 1_250.00
overdraft_limit  = -500.00
is_overdrawn     = balance < 0
can_withdraw     = balance > overdraft_limit

print(f"Authenticated: {is_authenticated}")
print(f"Can withdraw: {can_withdraw}")

# Truthiness -- any non-empty, non-zero value is truthy
pending_tasks = ["Deploy", "Test", "Review"]
if pending_tasks:
    print(f"{len(pending_tasks)} tasks still pending")
```

3.4 Type Conversion

Type conversion is the process of transforming a value from one data type to another. It arises constantly in practice: user input always arrives as a string; data read from a CSV file is a string until you convert it; API responses contain string representations of numbers that must be parsed before arithmetic. Understanding when and how to convert between types — and what can go wrong — is essential knowledge.

Implicit Conversion

Python performs implicit conversion (also called coercion) automatically in a small number of situations where the operation is unambiguous. The most common case is mixed integer/float arithmetic: when an int and a float are combined, Python promotes the int to a float so the result retains precision. Python does not silently coerce strings to numbers, however — that always requires an explicit call.

Example 1 — Introduction

Implicit int-to-float promotion in a mixed arithmetic expression.

```
unit_price = 12.50      # float
quantity   = 4          # int
discount   = 0.15       # float

# Python promotes quantity to float automatically
subtotal   = unit_price * quantity    # float: 50.0
saving     = subtotal * discount      # float: 7.5
total      = subtotal - saving        # float: 42.5

print(f"Subtotal: ${subtotal:.2f}")
print(f"Discount: -${saving:.2f}")
print(f"Total:    ${total:.2f}")
```

Explicit Conversion

Explicit conversion uses the built-in constructor functions: int(), float(), str(), and bool(). Each can fail with a ValueError if the input cannot be meaningfully converted — int("hello") raises an error, for example. Always wrap explicit conversions of external input in try/except to handle invalid data gracefully.

Example 2 — Variation

Explicit conversion of user input with error handling for invalid values.

```
def prompt_for_float(prompt):
    """Prompt the user until a valid float is entered."""
    while True:
        raw = input(prompt)
        try:
            return float(raw)
        except ValueError:
            print(f"  Invalid: {raw!r} is not a number. Try again.")

weight_kg = prompt_for_float("Enter package weight (kg): ")
length_cm = prompt_for_float("Enter length (cm): ")

volume_m3 = (length_cm / 100) ** 3
print(f"Weight: {weight_kg:.2f} kg, Volume: {volume_m3:.4f} m³")
```

Common Conversion Patterns

Example 3 — Real-world application

Type conversions encountered when parsing a CSV row from a sales report.

```
# Raw row from csv.DictReader -- all values are strings
row = {
    "order_id":   "ORD-20847",
    "quantity":   "12",
    "unit_price": "34.99",
```

```
    "is_rush":   "True",
    "discount":  "",
}

order_id    = row["order_id"]              # str -- keep as-is
quantity    = int(row["quantity"])         # "12"    -> 12
unit_price = float(row["unit_price"])     # "34.99" -> 34.99
is_rush     = row["is_rush"] == "True"     # "True"  -> True (bool)
discount    = float(row["discount"] or 0) # ""      -> 0.0

total = quantity * unit_price * (1 - discount)
print(f"Order {order_id}: {quantity} units @ ${unit_price:.2f} =
${total:.2f}")
```

3.5 The type() Function

type() returns the class of any object. Its primary uses are inspection during development and debugging — printing type(x) to understand what kind of value you are actually working with — and generating informative error messages. For production type-checking logic, isinstance() is almost always preferred because it handles subclasses correctly. A bool is an instance of int; type(True) == int returns False, but isinstance(True, int) returns True.

Inspection and Debugging

Example 1 — Introduction

Using type() to inspect values during debugging of a data pipeline stage.

```
def debug_pipeline_value(name, value):
    """Print the name, type, and repr of a pipeline value."""
    print(f"  {name}: type={type(value).__name__}, value={value!r}")

# Simulating inspection of values flowing through a data pipeline
raw_record = {"timestamp": "2024-07-15", "reading": "42.7", "valid": "1"}

debug_pipeline_value("timestamp", raw_record["timestamp"])
debug_pipeline_value("reading",   raw_record["reading"])
debug_pipeline_value("reading_as_float", float(raw_record["reading"]))
debug_pipeline_value("valid_as_bool",    bool(int(raw_record["valid"])))
```

isinstance() for Production Type Checks

When writing functions that need to enforce or adapt to input types, isinstance() is the right tool. It accepts a tuple of types as its second argument, allowing compact multi-type checks, and handles inheritance transparently:

Example 2 — Variation

isinstance() handling multiple acceptable numeric types in a statistics function.

```
def compute_statistics(values):
    """
    Compute mean and range for a list of numeric values.
    Accepts both int and float elements.
```

```
    """
    if not isinstance(values, list):
        raise TypeError(f"Expected list, got {type(values).__name__}")
    if not all(isinstance(v, (int, float)) for v in values):
        raise ValueError("All elements must be int or float")
    if not values:
        return None
    mean   = sum(values) / len(values)
    spread = max(values) - min(values)
    return {"mean": round(mean, 2), "range": round(spread, 2)}

print(compute_statistics([14.2, 18.7, 12.1, 19.4, 15.8]))
# {'mean': 16.04, 'range': 7.3}
```

3.6 Dynamic Typing in Python

Dynamic typing means that a variable's type is determined by the value it currently holds, not by a declaration made when the variable was created. In Python, you can reassign a variable to a value of a completely different type at any point. The interpreter tracks the type of the value, not the type of the name.

This is fundamentally different from statically typed languages. In Java, once you declare int count = 0, the variable count can only ever hold an int. In Python, count = 0 is simply a binding — you could later write count = "zero" and the interpreter would not object.

Flexibility and Its Trade-offs

Example 1 — Introduction

Dynamic typing enabling a configuration loader that handles values of any type.

```
# The same variable can hold different types at different points
config_value = "30"          # arrives from JSON as a string
config_value = int(config_value)  # converted to int for use
config_value = config_value * 1000  # now used as milliseconds

# This flexibility is valuable -- but can also conceal mistakes:
threshold = "100"  # accidentally left as string from earlier assignment
reading   = 87.3

# This silently fails to compare correctly:
# if reading > threshold:  # TypeError: > not supported between float and str
```

The flexibility of dynamic typing accelerates exploratory work and reduces boilerplate. The risk is that type errors surface at runtime rather than before the program runs. In a long-running production service, a type error triggered by an unusual input can cause a failure that would have been caught at compile time in a statically typed language.

Type Hints

Python 3.5 introduced optional type hints, which allow you to annotate variables and function signatures with their intended types. Type hints are not enforced at runtime — the interpreter ignores them during execution — but they serve as documentation and are checked by external tools like mypy and by IDE type checkers that can flag mismatches before you run the code.

Example 2 — Variation

Type hints on a function signature enabling IDE type checking and clear documentation.

```
from typing import Optional

def calculate_invoice_total(
    unit_price: float,
    quantity:   int,
    discount:   float = 0.0,
    tax_rate:   float = 0.10,
) -> float:
    """
    Calculate invoice total after discount and tax.
    Returns the final amount as a float rounded to 2 decimal places.
    """
    subtotal = unit_price * quantity
    after_discount = subtotal * (1 - discount)
    return round(after_discount * (1 + tax_rate), 2)

print(calculate_invoice_total(29.99, 5, discount=0.10))
# 149.95 (after 10% discount + 10% tax)
```

Type hints do not slow your program down. They make your intent explicit, allow tools to catch mistakes automatically, and make code significantly easier to understand for anyone who reads it. Adopting them consistently is considered good practice in professional Python work.

3.7 Using Variables in Expressions

An expression is any combination of values, variables, and operators that Python can evaluate to produce a result. Variables are the participants — expressions are where the computation actually happens. Understanding how Python evaluates expressions, the order in which it applies operators, and the rules for combining different types is fundamental to writing correct programs.

Arithmetic Expressions

Python supports the standard arithmetic operators: + (addition), - (subtraction), * (multiplication), / (true division, always returns float), // (floor division, returns int when both operands are int), % (modulo, returns remainder), and ** (exponentiation). Python follows standard mathematical precedence: parentheses first, then exponentiation, then multiplication/division/modulo, then addition/subtraction.

Example 1 — Introduction

Arithmetic expressions in a logistics cost calculation with operator precedence illustrated.

```
weight_kg      = 23.4
distance_km    = 512
base_rate      = 0.08       # dollars per kg per km
fuel_surcharge = 0.12       # 12% surcharge
insurance_rate = 0.005      # 0.5% of cargo value
cargo_value    = 3_800.00
```

```
# Operator precedence: * before +
freight_cost    = weight_kg * distance_km * base_rate
total_surcharge = freight_cost * fuel_surcharge
insurance       = cargo_value * insurance_rate

# Parentheses control grouping explicitly
grand_total = freight_cost + total_surcharge + insurance

print(f"Freight:    ${freight_cost:.2f}")
print(f"Surcharge:  ${total_surcharge:.2f}")
print(f"Insurance:  ${insurance:.2f}")
print(f"Total:      ${grand_total:.2f}")
```

Compound Assignment Operators

Python provides shorthand compound assignment operators that update a variable in place: += (add and assign), -= (subtract and assign), *= (multiply and assign), /= (divide and assign), //= (floor divide and assign), and %= (modulo and assign). These are cleaner and more readable than writing variable = variable + value:

```
running_total = 0.0
items = [14.99, 8.50, 23.00, 5.75, 11.20]

for price in items:
    running_total += price   # equivalent to: running_total = running_total +
price

print(f"Total: ${running_total:.2f}")  # Total: $63.44
```

Logical and Comparison Expressions

Comparison operators produce booleans: == (equal), != (not equal), < (less than), > (greater than), <= (less than or equal), >= (greater than or equal). Logical operators combine boolean expressions: and (both must be True), or (at least one must be True), not (inverts a boolean). Python also supports chained comparisons, which are more readable than their equivalent compound form.

Example 2 — Variation

Compound logical expressions in a loan eligibility check.

```
annual_income     = 85_000
credit_score      = 720
existing_debt_pct = 0.28   # debt-to-income ratio
loan_amount       = 450_000

INCOME_MIN        = 60_000
CREDIT_MIN        = 680
MAX_DTI           = 0.43
MAX_LTI           = 6.0     # loan-to-income ratio

# Chained comparison
```

```
income_ok   = annual_income >= INCOME_MIN
credit_ok   = credit_score  >= CREDIT_MIN
debt_ok     = existing_debt_pct <= MAX_DTI
lti_ok      = (loan_amount / annual_income) <= MAX_LTI

eligible = income_ok and credit_ok and debt_ok and lti_ok
print(f"Pre-approval: {'Yes' if eligible else 'No'}")
```

String Expressions

Example 3 — Real-world application

String operations building a formatted email notification from variable parts.

```
recipient_name  = "Jordan Patel"
order_id        = "ORD-20847"
item_count      = 3
dispatch_date   = "18 July 2024"
tracking_number = "AUP-7741928-X"

# f-string combining all variables into a formatted message
subject = f"Your order {order_id} has been dispatched"

body = (
    f"Dear {recipient_name},\n\n"
    f"Your order {order_id} containing {item_count} item(s) "
    f"was dispatched on {dispatch_date}.\n"
    f"Tracking number: {tracking_number}\n\n"
    "Thank you for your order."
)

print(subject)
print(body)
```

Chapter 3 Summary

- A variable is a named binding to a value in memory. Python is dynamically typed: variables do not carry a declared type, and the same name can point to different types as a program runs.
- Use type() to inspect the type of a value during debugging. Use isinstance() for type-checking in production code — it handles subclasses correctly and accepts tuples of types for multi-type checks.
- Name variables using snake_case, constants using UPPER_CASE, and classes using PascalCase, following PEP 8. Names should describe purpose, not structure. Ambiguous names like data, temp, and x impose a reading cost on everyone who encounters them.
- The four fundamental types: int (whole numbers, arbitrary precision), float (decimal numbers, IEEE 754 double precision), str (immutable sequence of characters), bool (True or False, subclass of int).
- Use explicit conversion functions — int(), float(), str(), bool() — to change types. Wrap conversions of external input in try/except to handle invalid values gracefully. Python

performs implicit int-to-float promotion in mixed arithmetic but never silently converts strings to numbers.

- Dynamic typing gives flexibility and reduces boilerplate but moves type errors to runtime. Optional type hints (PEP 484, available from Python 3.5) document intended types and enable static analysis tools to catch mismatches before the program runs.
- Expressions combine variables, literals, and operators to produce values. Python follows standard operator precedence; use parentheses to make order explicit. Compound assignment operators (+=, -=, *= etc.) update variables cleanly. Chained comparisons (0 < x < 100) are valid and more readable than their compound equivalents.

Guided Exercises

Complete the following exercises in VS Code, IDLE, or your preferred Python environment. Each exercise targets a specific concept from this chapter.

1. Write a shipping rate calculator script. Define at least five constants at the top of the file (base rate, express surcharge, fragile surcharge, free-shipping threshold, GST rate). Write a function calculate_shipping(weight_kg, order_total, express=False, fragile=False) that uses those constants and returns the total shipping cost. Test it with at least four different combinations of arguments and print a formatted result for each.

2. Write a function parse_sales_record(row) that accepts a dictionary with string values (as they would come from a CSV file) and returns a new dictionary with correctly typed values: order_id as str, quantity as int, unit_price as float, is_priority as bool, and discount as float (defaulting to 0.0 if the value is empty). Demonstrate it with at least two test rows, one of which has an empty discount field.

3. Write a function compute_statistics(values) that accepts a list, validates that every element is an int or float using isinstance(), and returns a dictionary containing mean, minimum, maximum, and range. Raise a TypeError with a descriptive message if any element is not numeric. Test it with valid input, and show what happens with invalid input using try/except.

4. Rewrite the following function using type hints on all parameters and the return value, then add a complete docstring. Original: def calc(p, q, d, t): return round(p * q * (1 - d) * (1 + t), 2). Use meaningful parameter names. Demonstrate that the function still works correctly with at least three test calls.

5. Write a script that builds a formatted dispatch notification email (subject line and body) from at least six variables: recipient name, order ID, item count, total value, dispatch date, and tracking number. Use f-strings throughout. Then demonstrate how changing one variable (e.g. the recipient name or order total) automatically flows through to the formatted output.

Challenge Tasks

The following tasks are open-ended. No hints are provided. They require you to design a solution, not just apply a pattern.

1. Design and implement a payroll calculator. It should: define all rate constants (hourly rate, overtime multiplier, tax brackets with thresholds, superannuation rate) at the top of the file; accept hours_worked and employee_id as inputs; compute regular pay, overtime pay (hours above 38 at 1.5x), gross pay, tax (applying the correct bracket), super contribution, and net pay; and print a formatted pay slip. Use meaningful variable names and type hints throughout.

2. Write a robust CSV row parser that handles a realistic set of data quality issues. Given a list of raw string dictionaries (simulating csv.DictReader output), your parser should: convert each field to the correct type; handle empty strings by substituting sensible defaults; catch and log rows that contain values that cannot be converted (e.g. non-numeric in a price field); and return two lists: successfully parsed records and failed records with an error description. Test with at least eight rows, including at least two that contain intentional errors.

3. Build a configuration validator for a hypothetical web application. The config is a Python dictionary with keys for host (str), port (int, must be 1–65535), debug (bool), max_connections (int, must be positive), timeout_seconds (float, must be positive), and allowed_origins (list of strings). Write a validate_config(config) function that checks every field for the correct type and valid range, collects all errors (not just the first one), and raises a ValueError listing all issues if any are found. Demonstrate it with a valid config and at least two configs containing different combinations of errors.

CHAPTER 4 User Input and Output

Think, Code, Create – A Python Programming Journey

> A program that cannot receive information from the outside world and cannot communicate results back is useful only in the narrowest of contexts. The ability to accept input and produce meaningful output is what transforms a static script into a tool that responds, adapts, and delivers value. This chapter covers the full picture of interactive I/O in Python: reading from the terminal with input(), writing to standard output and error with print(), handling the type conversion that terminal input always requires, formatting output with f-strings, and validating user-supplied data robustly.
>
> Section 4.7 brings these skills together into a systematic approach to input validation — one of the most practically important techniques in this book. By the end of this chapter you will be able to write interactive command-line programs that collect input cleanly, convert it correctly, display results precisely, and handle every foreseeable error without crashing.

4.1 Input and Output in Programs

Input is any information a program receives from the outside world. Output is any information a program sends back. Every useful program does both — they are the interface between the logic inside the program and the environment it operates in.

Terminal I/O, which this chapter focuses on, is only one form. Programs also receive input from files, databases, network sockets, environment variables, and command-line arguments; and they produce output to files, APIs, databases, and logging systems. The principles you learn here apply to all of them. The input() function and print() function are the simplest, most immediate examples of a pattern that recurs at every level of software development.

Why I/O Design Matters

The quality of a program's I/O has a direct impact on how reliably it works and how easy it is to use. Poorly designed I/O fails silently on invalid input, produces output that is hard to read or parse, and mixes error messages with normal output in ways that break downstream consumers. Well-designed I/O validates input before using it, formats output consistently, separates normal output from error messages, and provides clear feedback when something goes wrong.

These are not cosmetic concerns. A script that crashes with an unhandled ValueError when someone enters a space in a numeric field is a broken script, regardless of how correct the core logic is. A reporting script that writes error messages into the same stream as its data output will corrupt any automated pipeline consuming that data. I/O discipline is engineering discipline.

The Two Standard Streams

Every process running on a modern operating system has at least two output streams: standard output (stdout) and standard error (stderr). By convention, normal program output goes to stdout and diagnostic or error messages go to stderr. This separation allows the two streams to be handled independently — redirected to different files, piped to different processes, or filtered separately. Python's print() writes to stdout by default; passing file=sys.stderr directs output to stderr instead.

Example 1 — Introduction

A data processing script that correctly separates normal output from error messages across the two standard streams.

```
import sys

def process_record(record_id: str, value: str):
    try:
        amount = float(value)
        print(f"Processed {record_id}: ${amount:.2f}")   # stdout
        return amount
    except ValueError:
        print(f"ERROR: {record_id} has invalid value: {value!r}",
              file=sys.stderr)                            # stderr
        return None

records = [("TXN-001", "149.95"), ("TXN-002", "N/A"), ("TXN-003", "87.50")]
totals  = [process_record(r, v) for r, v in records]
valid   = [t for t in totals if t is not None]
print(f"\nTotal of {len(valid)} valid records: ${sum(valid):.2f}")
```

4.2 Using input() and print()

input() and print() are the two functions at the centre of this chapter. input() suspends execution, displays an optional prompt, reads a line of text from the keyboard, and returns it as a string. print() evaluates its arguments, converts them to strings, and writes them to stdout followed by a newline. Both are deceptively simple to use and have important nuances worth understanding thoroughly.

input()

The prompt argument to input() is optional but almost always used — a program that waits silently for input is confusing. The returned value is always a str, regardless of what the user types. If the user presses Enter with nothing typed, the return value is an empty string. If the user closes the input stream (Ctrl+D on Unix, Ctrl+Z on Windows), input() raises an EOFError — worth handling in scripts that may be run non-interactively.

Example 1 — Introduction

Collecting multiple fields of input for a new customer record, with basic prompting.

```
print("=== New Customer Registration ===")

full_name = input("Full name:          ")
email     = input("Email address:      ")
```

```
phone     = input("Phone (optional):   ")

print("\n--- Record to be created ---")
print(f"  Name:  {full_name}")
print(f"  Email: {email}")
if phone.strip():
    print(f"  Phone: {phone.strip()}")
else:
    print("  Phone: not provided")
```

print()

print() accepts any number of positional arguments, converts each to a string via str(), and writes them separated by sep (default: a single space) and followed by end (default: a newline). The flush parameter forces the output buffer to be written immediately, which matters in real-time progress reporting where output would otherwise be held until the buffer fills.

Example 2 — Variation

print() parameters: sep, end, and flush demonstrated in a progress reporting context.

```
import time, sys

# sep: custom separator between arguments
print("host", "port", "status", sep=" | ")
# host | port | status

# end + flush: real-time progress on a single line
stages = ["Connecting", "Authenticating", "Downloading", "Verifying", "Done"]
for stage in stages:
    print(f"\r[{stage:<16}]", end="", flush=True)
    time.sleep(0.4)
print()  # final newline after the loop

# Structured log line using sep
print("2024-07-15T09:14:02", "INFO", "Service started",
      sep="  |  ", file=sys.stdout)
```

Input Always Returns a String

This is the single most common source of errors for developers new to Python I/O. Every value from input() is a str. Adding two inputs without conversion produces concatenation, not addition. Comparing an input value to a number with == will never match. The rule is absolute: always convert input before using it in arithmetic, comparisons, or any operation that requires a specific type.

Example 3 — Real-world application

A unit converter demonstrating the correct pattern: collect input as str, convert immediately, then compute.

```
print("=== Unit Converter: kg → pounds ===")

raw    = input("Enter weight in kilograms: ")
kg     = float(raw)            # convert before arithmetic
```

```
pounds = kg * 2.20462

print(f"{kg:.3f} kg  =  {pounds:.3f} lb")

# Common mistake -- without conversion:
# raw1 = input("First:  ")
# raw2 = input("Second: ")
# print(raw1 + raw2)   -- prints "12" + "34" = "1234", not 46
```

4.3 String Concatenation with Input

String concatenation is the joining of two or more strings to form a single string. In the context of input handling, concatenation is how you build formatted output from a mixture of fixed text and user-supplied values. Python offers three main approaches: the + operator, the join() method, and f-strings. Understanding when to use each produces cleaner, more readable code.

The + Operator

The + operator joins two strings directly. It requires all operands to be strings — mixing a string and a number raises a TypeError. Any non-string value must be explicitly converted with str() first. For short constructions with one or two variables, + is clear and direct.

Example 1 — Introduction

Building a structured filename from components using the + operator.

```
report_type = input("Report type (daily/weekly/monthly): ").strip().lower()
region      = input("Region code (e.g. AU-NSW):           ").strip().upper()
date_stamp  = "2024-07-15"

filename = report_type + "_" + region + "_" + date_stamp + ".csv"
print("Output file: " + filename)
# Output file: daily_AU-NSW_2024-07-15.csv
```

The join() Method

str.join(iterable) concatenates all elements of an iterable of strings, placing the str between each pair. It is the correct approach when combining a list of items — more efficient than repeated + because it avoids creating intermediate string objects for each step.

Example 2 — Variation

Collecting multiple items into a list then joining them for display and file output.

```
print("Enter line items (press Enter after each; blank line to finish):")

items = []
while True:
    item = input(f"  Item {len(items)+1}: ").strip()
    if not item:
        break
    items.append(item)

# join for display
```

```
print("\nItems: " + ", ".join(items))

# join for CSV output
csv_row = "|".join(items)
print("CSV:    " + csv_row)
```

f-strings (Preferred for Mixed Output)

For any output that mixes fixed text with variable values — which is the majority of output in practice — f-strings are cleaner than concatenation. They avoid the need for explicit str() calls, support format specifiers inline, and remain readable as the number of variables grows. The + operator becomes increasingly awkward beyond two variables; f-strings do not.

Example 3 — Real-world application

A formatted shipping label built from multiple input fields using f-string composition.

```
print("=== Parcel Dispatch Entry ===")

sender     = input("Sender name:       ")
recipient  = input("Recipient name:    ")
suburb     = input("Destination suburb: ")
postcode   = input("Postcode:           ")
weight_kg  = float(input("Weight (kg):       "))

label = (
    f"FROM:  {sender}\n"
    f"TO:    {recipient}\n"
    f"DEST:  {suburb.upper()}  {postcode}\n"
    f"MASS:  {weight_kg:.2f} kg"
)

print("\n" + "="*30)
print(label)
print("="*30)
```

4.4 Typecasting Input Values

Because input() always returns a string, typecasting — converting the string to the appropriate type before use — is required whenever the value needs to participate in arithmetic, comparison, or any operation specific to a non-string type. The conversion functions int(), float(), bool(), and their inverse str() handle the most common cases. Failing to convert produces either a TypeError or a silently wrong result.

int() and float()

int() converts a string to an integer. It accepts strings containing only digits (with an optional leading minus sign) and raises ValueError for anything else. float() is more permissive — it accepts integers, decimal strings, and scientific notation — and is the right choice when fractional values are possible. int() truncates rather than rounds: int(3.9) is 3.

Example 1 — Introduction

A bulk pricing calculator that casts integer quantity and float unit price from input.

```
print("=== Bulk Order Pricing ===")

product    = input("Product name:      ")
quantity   = int(input("Quantity:          "))
unit_price = float(input("Unit price ($):   "))
discount   = float(input("Discount (%):     ")) / 100

subtotal   = quantity * unit_price
saving     = subtotal * discount
total      = subtotal - saving

print(f"\n  {product}")
print(f"  {quantity} × ${unit_price:.2f} = ${subtotal:.2f}")
print(f"  Discount ({discount*100:.0f}%):  -${saving:.2f}")
print(f"  Total:             ${total:.2f}")
```

bool() and Truthy Input

bool() converts any value to True or False according to Python's truthiness rules. For user input, however, bool() is almost never what you want — bool("False") returns True because any non-empty string is truthy. To convert a user's yes/no or true/false response to a boolean, compare the normalised string explicitly:

```
raw    = input("Send confirmation email? (yes/no): ").strip().lower()
notify = raw in ("yes", "y", "1", "true")
print(f"Notification: {'enabled' if notify else 'disabled'}")
```

Chaining Conversion and Input

It is common to wrap input() and the conversion call on a single line. This is clear and concise when the input prompt makes the expected type obvious and when you are already handling conversion errors at a higher level in the program:

Example 2 — Variation

Chained conversion with a validated re-entry loop for a measurement entry tool.

```
def get_float(prompt: str, min_val: float = None, max_val: float = None) ->
float:
    """Prompt until the user enters a valid float within optional bounds."""
    while True:
        try:
            value = float(input(prompt))
        except ValueError:
            print("  Please enter a numeric value.")
            continue
        if min_val is not None and value < min_val:
            print(f"  Value must be at least {min_val}.")
        elif max_val is not None and value > max_val:
            print(f"  Value must be at most {max_val}.")
        else:
            return value
```

```
length = get_float("Length (m, 0.1-100): ", 0.1, 100)
width  = get_float("Width  (m, 0.1-100): ", 0.1, 100)
area   = length * width
print(f"Floor area: {area:.2f} m²")
```

4.5 Creating Interactive Scripts

An interactive script is one that maintains a dialogue with the user: it prompts, receives input, responds, and either prompts again or exits based on the user's choices. The typical structure is a while True loop containing a menu, a dispatch block that routes to the appropriate action, and an exit condition. This pattern — sometimes called a REPL (read-evaluate-print loop) in its most general form — is the backbone of every command-line tool.

The Menu Loop Pattern

A clean menu loop has three components: a display function that shows the options, a dispatch block (if/elif/else or a dictionary) that calls the correct handler, and a sentinel value that breaks the loop. The loop continues until the sentinel is entered; every other input routes to a handler or to a "not recognised" fallback.

Example 1 — Introduction

A contact manager with add, search, display, and exit operations — the canonical menu-loop pattern.

```
contacts = {}

def add_contact():
    name  = input("  Name:  ").strip()
    email = input("  Email: ").strip()
    if name:
        contacts[name.lower()] = email
        print(f"  Added: {name}")

def search_contact():
    term = input("  Search: ").strip().lower()
    found = {k: v for k, v in contacts.items() if term in k}
    if found:
        for name, email in found.items():
            print(f"  {name.title()}: {email}")
    else:
        print("  No matches found.")

while True:
    print("\n[1] Add  [2] Search  [3] List all  [Q] Quit")
    choice = input("> ").strip().lower()
    if   choice == "1": add_contact()
    elif choice == "2": search_contact()
    elif choice == "3":
        for n, e in contacts.items(): print(f"  {n.title()}: {e}")
    elif choice == "q": print("Goodbye."); break
    else: print("  Unrecognised option.")
```

Multi-Step Data Collection

Not every interactive script is a menu loop. Some collect a structured set of inputs once, process them, and exit. This pattern is common for data-entry tools, configuration wizards, and report generators. The key discipline is separating the collection phase from the processing phase — gather all inputs first, validate them, then compute and display results.

Example 2 — Variation

A loan repayment calculator that collects inputs, validates them, then displays a full amortisation summary.

```
print("=== Loan Repayment Calculator ===")

principal    = float(input("Loan amount ($):        "))
annual_rate  = float(input("Annual interest rate (%):")) / 100
term_years   = int(input("Loan term (years):       "))

monthly_rate = annual_rate / 12
n_payments   = term_years * 12

if monthly_rate == 0:
    monthly_payment = principal / n_payments
else:
    monthly_payment = (principal * monthly_rate *
                       (1 + monthly_rate) ** n_payments /
                       ((1 + monthly_rate) ** n_payments - 1))

total_paid     = monthly_payment * n_payments
total_interest = total_paid - principal

print(f"\n  Monthly payment: ${monthly_payment:,.2f}")
print(f"  Total repaid:    ${total_paid:,.2f}")
print(f"  Interest paid:   ${total_interest:,.2f}")
```

Looping Until Valid

A common sub-pattern within interactive scripts is looping until the user provides valid input for a specific field. Encapsulate this in a dedicated input function so the main script flow remains clean and readable. The function handles all prompting, conversion, and re-prompting internally; the caller simply receives a valid value.

Example 3 — Real-world application

A stock entry tool built from reusable validated-input helpers.

```
def get_string(prompt: str, min_length: int = 1) -> str:
    while True:
        val = input(prompt).strip()
        if len(val) >= min_length:
            return val
        print(f"  Must be at least {min_length} character(s).")

def get_positive_int(prompt: str) -> int:
    while True:
        try:
            val = int(input(prompt))
```

```python
            if val > 0: return val
            print("  Must be a positive integer.")
        except ValueError:
            print("  Please enter a whole number.")

def get_positive_float(prompt: str) -> float:
    while True:
        try:
            val = float(input(prompt))
            if val > 0: return val
            print("  Must be greater than zero.")
        except ValueError:
            print("  Please enter a numeric value.")

sku      = get_string("SKU:              ", min_length=4)
quantity = get_positive_int("Quantity received: ")
cost     = get_positive_float("Unit cost ($):     ")
print(f"  Stocked {quantity} x {sku} @ ${cost:.2f} = ${quantity*cost:.2f}")
```

4.6 Formatting Strings with f-strings

f-strings (formatted string literals) embed Python expressions directly into string literals and evaluate them at runtime. Introduced in Python 3.6, they are the recommended way to produce formatted output in all modern Python code. Their syntax is concise, their performance is excellent, and they eliminate the awkward str() conversion calls and + operators that accumulate when building strings by concatenation.

Format Specifiers

An expression inside an f-string curly brace can include a format specifier after a colon. The specifier controls how the value is presented: its width, alignment, fill character, numeric precision, and notation. The most commonly used specifiers are worth memorising.

Example 1 — Introduction

A financial report formatted entirely with f-string specifiers: currency, percentages, alignment, separators.

```python
accounts = [
    ("Operating Account",  1_284_750.42,  12.4),
    ("Reserve Fund",       3_500_000.00,   8.1),
    ("Payroll Holding",      187_230.15,  31.7),
]

print(f"{"Account":<22} {"Balance":>14} {"Change":>8}")
print("-" * 46)
for name, balance, change in accounts:
    arrow = "▲" if change >= 0 else "▼"
    print(f"{name:<22} ${balance:>13,.2f} {arrow}{abs(change):>5.1f}%")
```

Key specifiers used above: :<22 (left-align in 22 chars), :>14 (right-align in 14 chars), :>13,.2f (right-align, thousands separator, 2 decimal places), :>5.1f (right-align, 1 decimal place, min 5 chars wide).

Expressions and Conversions Inside f-strings

The expression inside the curly braces can be any valid Python expression: arithmetic, method calls, conditional expressions, even function calls. Two special conversion flags are also available: !r applies repr() (useful for debugging, adds quotes around strings), and !s applies str() (the default).

Example 2 — **Variation**

f-strings with inline expressions, conditional logic, and the !r conversion flag.

```
status_code = 404
endpoint    = "/api/users/99"
response_ms = 3.7

# Inline conditional expression
severity = "WARN" if status_code < 500 else "ERROR"
print(f"[{severity}] {status_code} {endpoint} ({response_ms:.1f}ms)")

# !r for unambiguous representation of string values
raw_input = "  42abc  "
print(f"Raw input received: {raw_input!r}")
# Raw input received: '  42abc  '

# Inline method call
city = "  darwin  "
print(f"City: {city.strip().title()}")
# City: Darwin
```

Debugging with f-strings

Python 3.8 introduced the = specifier inside f-string expressions. Writing {expression=} prints both the expression text and its value, making it a compact debugging tool that is faster to write than a full print(f"variable = {variable}") statement and clearer about which variable is being reported.

Example 3 — **Real-world application**

The = specifier for rapid debugging of intermediate values in a multi-step calculation.

```
units       = 47
unit_cost   = 12.50
discount    = 0.15
tax_rate    = 0.10

subtotal    = units * unit_cost
after_disc  = subtotal * (1 - discount)
final       = after_disc * (1 + tax_rate)

# = specifier: prints "subtotal = 587.5" etc.
print(f"{subtotal=:.2f}  {after_disc=:.2f}  {final=:.2f}")
# subtotal=587.50  after_disc=499.38  final=549.31
```

4.7 Input Validation and Error Handling

Input validation is the practice of checking that a value meets all requirements before using it in the program's logic. It is not optional. Any input that arrives from outside the program — terminal input, file contents, API responses, database records — must be treated as potentially wrong, in the wrong format, or missing entirely. Programs that skip validation eventually encounter unexpected input and crash or, worse, produce silently incorrect results.

The Validation Loop

The foundational pattern for interactive input validation is a while True loop that prompts for a value, attempts the conversion and validation, returns the value on success, and prints a helpful message and continues the loop on failure. Wrapping this in a reusable function keeps the calling code clean:

Example 1 — Introduction

Reusable validated-input functions for integers and floats with range constraints.

```
def prompt_int(prompt: str, min_val: int = None, max_val: int = None) -> int:
    while True:
        try:
            val = int(input(prompt))
        except ValueError:
            print("  Please enter a whole number.")
            continue
        if min_val is not None and val < min_val:
            print(f"  Must be at least {min_val}.")
        elif max_val is not None and val > max_val:
            print(f"  Must be at most {max_val}.")
        else:
            return val

employees = prompt_int("Number of employees (1-10000): ", 1, 10_000)
print(f"Processing payroll for {employees:,} employees.")
```

Empty Input and String Validation

Empty input — the user pressing Enter with nothing typed — returns an empty string. Always call .strip() before checking length or content, because a string containing only spaces is functionally empty. Validate string fields for minimum length, allowed characters, or required formats as appropriate for the domain.

Example 2 — Variation

Validating a non-empty string field and a date string in a required format.

```
from datetime import datetime

def prompt_nonempty(prompt: str) -> str:
    while True:
        val = input(prompt).strip()
        if val:
            return val
        print("  This field cannot be empty.")
```

```
def prompt_date(prompt: str, fmt: str = "%d/%m/%Y") -> datetime:
    while True:
        raw = input(prompt).strip()
        try:
            return datetime.strptime(raw, fmt)
        except ValueError:
            print(f"  Please use the format {fmt} (e.g. 15/07/2024).")

client_name  = prompt_nonempty("Client name:         ")
invoice_date = prompt_date("Invoice date (DD/MM/YYYY): ")
due_date     = prompt_date("Payment due (DD/MM/YYYY):  ")

days_to_pay  = (due_date - invoice_date).days
print(f"  Net {days_to_pay} days for {client_name}")
```

Collecting Multiple Values with Error Summary

In batch data entry or form processing, rather than halting at the first error, it is often better to collect all inputs and report all validation failures together. This gives the user a complete picture of what needs correcting.

Example 3 — Real-world application

A product entry form that collects all fields, validates them, and reports all errors before accepting the record.

```
def validate_product(sku, name, price_str, stock_str):
    errors = []
    if len(sku.strip()) < 4:
        errors.append("SKU must be at least 4 characters.")
    if not name.strip():
        errors.append("Product name is required.")
    try:
        price = float(price_str)
        if price <= 0: errors.append("Price must be greater than zero.")
    except ValueError:
        errors.append(f"Invalid price: {price_str!r}")
        price = None
    try:
        stock = int(stock_str)
        if stock < 0: errors.append("Stock cannot be negative.")
    except ValueError:
        errors.append(f"Invalid stock quantity: {stock_str!r}")
        stock = None
    return errors, price, stock

sku   = input("SKU:      ").strip()
name  = input("Name:     ").strip()
price = input("Price:    ").strip()
stock = input("Stock:    ").strip()

errors, price_val, stock_val = validate_product(sku, name, price, stock)
if errors:
    print("\nValidation failed:")
    for err in errors: print(f"  • {err}")
else:
```

```
    print(f"\nAccepted: {sku} | {name} | ${price_val:.2f} | {stock_val}
units")
```

Chapter 4 Summary

- Every process has two standard output streams: stdout for normal output, stderr for errors and diagnostics. print() writes to stdout by default; print(..., file=sys.stderr) writes to stderr. Separating them allows pipelines and tools to handle each stream independently.
- input() suspends execution, displays an optional prompt, and returns the user's input as a str. The return type is always str — no exceptions. Arithmetic, comparisons, and any type-specific operations require explicit conversion before use.
- String concatenation with + requires all operands to be str; non-strings must be converted first. str.join(iterable) is more efficient than repeated + for combining lists of strings. f-strings are the preferred approach whenever output mixes fixed text with variable values.
- Typecasting: int() converts to integer (truncates, not rounds); float() accepts integers, decimals, and scientific notation; bool() on a non-empty string is always True — compare the stripped string to expected values explicitly for yes/no input.
- Interactive scripts follow a while True menu loop: display options, read choice, dispatch to handler, break on the exit sentinel. Separate input collection from processing logic by encapsulating validation in dedicated functions that return only valid values.
- f-string format specifiers control width (number), alignment (< left, > right, ^ centre), fill, thousands separator (,), decimal precision (.Nf), and notation. The = specifier (Python 3.8+) prints both the expression and its value, making it a quick debugging tool.
- Input validation is mandatory for any input from outside the program. The validation loop pattern: try the conversion inside while True, handle the exception with a helpful message and continue, return only on success. For forms, collect all errors before reporting rather than halting at the first one.

Guided Exercises

Complete the following exercises in VS Code, IDLE, or your preferred Python environment. Each exercise targets a specific concept from this chapter.

1. Write a script that collects the details for a new employee record: full name (non-empty string), department (must be one of: Engineering, Finance, Operations, HR), annual salary (positive float), and start date (DD/MM/YYYY format). Use separate validated-input helper functions for each field type. Print a formatted summary of the accepted record.

2. Build a menu-driven currency converter. Store at least five exchange rates in a dictionary at the top of the file. The menu should offer: (1) USD to other currency, (2) other currency to USD, (3) show all rates, (Q) quit. Use a while True loop and route each choice to a handler function. Validate that any amount entered is a positive float before converting.

3. Write a function format_report_row(label, value, unit, width=20) that returns a formatted string with the label left-aligned in a field of width characters, the value right-aligned in 12 characters with two decimal places, and the unit appended. Use it to print at least five rows of a systems report (e.g. CPU usage, memory used, disk read rate) with a header and separator line.

4. Create a product entry form that collects SKU, name, price, and quantity for up to ten products. After each entry, validate all four fields (SKU min 4 chars, name non-empty, price positive float, quantity non-negative int) and either accept the record or print all validation errors. When the user enters a blank SKU, exit the loop and print a summary of total products entered and total inventory value.

5. Demonstrate the f-string = specifier and at least four different format specifiers in a single script. Write a compound interest calculator that accepts principal, annual rate, and term, then uses the = specifier to show intermediate values during debugging, and uses alignment and precision specifiers to produce a clean final report.

Challenge Tasks

The following tasks are open-ended. No hints are provided. They require you to design a solution, not just apply a pattern.

1. Design and build a complete command-line invoice generator. It should: prompt for client name and a line of items (description, quantity, unit price) until a blank description ends entry; validate every numeric field; calculate subtotal, GST (10%), and total; and print a formatted invoice with aligned columns, a separator line, and all monetary values formatted to two decimal places. The invoice should also display the number of items and the largest single line item.

2. Write a robust input collection library as a module (inputs.py) containing at least six reusable functions: prompt_string (non-empty, optional min/max length), prompt_int (optional min/max), prompt_float (optional min/max), prompt_bool (yes/no), prompt_choice (must be one of a provided list), and prompt_date (configurable format). Each function must loop until valid input is received and print specific, actionable error messages. Write a second script (demo.py) that imports and demonstrates all six functions with realistic prompts.

3. Build a payslip generator with full input validation and formatted output. The program should collect: employee name, employee ID, hourly rate (float, > 0), hours worked this period (float, >= 0), and overtime hours (float, >= 0, cannot exceed total hours). Calculate: regular pay (up to 38h at normal rate), overtime pay (above 38h at 1.5x, or use the overtime input if hours_worked <= 38), gross pay, tax (use a simple bracket: 0-18200 tax-free, 18201-45000 at 19%,

45001-120000 at 32.5%, >120000 at 37% — applied to the annualised gross), super (11.5% of gross), and net pay. Print a formatted payslip with every line item clearly labelled and all values right-aligned.

CHAPTER 5 **Arithmetic and Expressions**

Think, Code, Create – A Python Programming Journey

> Arithmetic is the engine beneath almost every useful program. Financial calculations, scientific modelling, data analysis, logistics optimisation, and engineering simulation all reduce to sequences of arithmetic operations on variables. Python provides a complete set of arithmetic operators, a well-defined precedence system, a rich standard library of mathematical functions, and robust tools for managing numeric precision. This chapter covers all of them.
>
> Sections 5.6 and 5.7 are particularly important: practical problem-solving shows how the operators and functions combine to solve real computational tasks, while the common mistakes section addresses the floating-point precision issues that catch out even experienced developers. By the end of this chapter you will be able to write accurate, well-structured arithmetic expressions across a wide range of applied contexts.

5.1 Arithmetic Operators

Python provides seven arithmetic operators that cover all standard mathematical operations. Understanding what each one does, what type it returns, and where the edge cases lie is essential before combining them in expressions.

Addition, Subtraction, and Negation

+ adds two numeric values. - subtracts the right operand from the left, or negates a single value when used as a unary operator. Both work with int and float operands; the result is float if either operand is float, int if both are int.

Example 1 — Introduction

Addition and subtraction in a warehouse stock reconciliation script.

```
opening_stock  = 2_840
received       = 512
dispatched     = 673
damaged        = 14

closing_stock  = opening_stock + received - dispatched - damaged
variance       = closing_stock - 2_665   # expected count

print(f"Closing stock:  {closing_stock:>6,}")
print(f"Variance:       {variance:>+6,}")   # + forces sign display
```

Multiplication and Division

* multiplies two values and returns the same type rules as addition. / (true division) always returns a float, even when both operands are integers and the result is mathematically whole. //

(floor division) returns the largest integer less than or equal to the true quotient — when both operands are int the result is int; if either is float the result is float. ** raises the left operand to the power of the right.

Example 2 — Variation

Division operators contrasted in a packing and dispatch context.

```
units_to_pack = 1_450
units_per_box = 24

true_div    = units_to_pack / units_per_box    # 60.41… (float)
full_boxes = units_to_pack // units_per_box    # 60 (int)
loose       = units_to_pack %  units_per_box    # 10 (int)

print(f"True division:  {true_div:.4f}")
print(f"Full boxes:     {full_boxes}")
print(f"Loose units:    {loose}")

# Verification: floor division and modulus satisfy the identity
# a == (a // b) * b + (a % b)
assert units_to_pack == full_boxes * units_per_box + loose
```

Modulus (%) and Exponentiation (**)

% returns the remainder after floor division. Its sign follows the divisor, not the dividend: -7 % 3 is 2 (not -1), because Python defines modulus so that a == (a // b) * b + (a % b) holds for all integers. This makes % useful for cyclic patterns and determining divisibility, but requires care with negative values.

** raises the left operand to the power of the right. For large integer exponents the result is an exact integer; for float exponents it uses logarithms and may introduce small floating-point errors.

Example 3 — Real-world application

Modulus for round-robin task assignment and exponentiation for compound growth modelling.

```
# Modulus: distribute tasks across workers in round-robin order
WORKERS = ["Alice", "Bob", "Carlos", "Diana"]
tasks   = ["Task A", "Task B", "Task C", "Task D", "Task E", "Task F"]

for i, task in enumerate(tasks):
    assigned_to = WORKERS[i % len(WORKERS)]
    print(f"  {task}  →  {assigned_to}")

# Exponentiation: compound annual growth
revenue_2020    = 4_200_000
annual_growth   = 0.087   # 8.7% CAGR
years           = 5
revenue_2025    = revenue_2020 * (1 + annual_growth) ** years
print(f"\nProjected 2025 revenue: ${revenue_2025:,.0f}")
```

5.2 Operator Precedence

When an expression contains multiple operators, Python evaluates them in a defined order called operator precedence. The rules mirror standard mathematical convention. From highest to lowest: parentheses () → exponentiation ** → unary + and - → multiplication *, true division /, floor division //, and modulus % → addition + and subtraction -. Comparison and logical operators are lower still and are covered in Chapter 6.

Operators at the same precedence level are evaluated left to right (left-associative), with one notable exception: ** is right-associative, so 2 ** 3 ** 2 evaluates as 2 ** (3 ** 2) = 2 ** 9 = 512, not (2 ** 3) ** 2 = 64.

Precedence in Practice

Example 1 — Introduction

A payroll expression demonstrating why precedence matters — and why parentheses are the clearest solution.

```
hourly_rate    = 38.50
hours_regular = 38
hours_overtime = 4
OT_MULTIPLIER  = 1.5

# WRONG: * has higher precedence than +, so this computes
# hourly_rate + (hours_overtime * OT_MULTIPLIER * hourly_rate)
wrong = hourly_rate + hours_overtime * OT_MULTIPLIER * hourly_rate
print(f"Wrong result: ${wrong:,.2f}")

# CORRECT: parentheses make intent explicit
regular_pay  = hours_regular  * hourly_rate
overtime_pay = hours_overtime * hourly_rate * OT_MULTIPLIER
gross_pay    = regular_pay + overtime_pay
print(f"Gross pay:    ${gross_pay:,.2f}")
```

Using Parentheses for Clarity

Even when parentheses are not strictly required by precedence, use them whenever an expression involves more than two operators. A reader should not need to mentally apply the precedence table to understand what a line of code computes. Parentheses are documentation as much as they are syntax.

Example 2 — Variation

Multi-operator financial expression: parentheses make the calculation unambiguous.

```
principal     = 250_000
annual_rate   = 0.068
monthly_rate  = annual_rate / 12
n_payments    = 30 * 12    # 30-year term

# Mortgage payment formula -- parentheses are essential here
payment = (
    principal
    * (monthly_rate * (1 + monthly_rate) ** n_payments)
    / ((1 + monthly_rate) ** n_payments - 1)
```

```
)
print(f"Monthly payment: ${payment:,.2f}")
```

Right-Associativity of **

Example 3 — Real-world application

*Demonstrating ** right-associativity and how it affects compound exponent expressions.*

```
# ** is right-associative: a ** b ** c == a ** (b ** c)
print(2 ** 3 ** 2)    # 2 ** (3**2) = 2**9 = 512
print((2 ** 3) ** 2) # (2**3)**2  = 8**2  = 64

# In engineering: stress concentration factor calculation
# K_t = 1 + C * (r/d) ** n
C, r, d, n = 2.0, 0.5, 10.0, 0.65
K_t = 1 + C * (r / d) ** n
print(f"Stress concentration factor: {K_t:.4f}")
```

5.3 Expressions and Assignment

An expression is any syntactically valid combination of values, variables, operators, and function calls that Python can evaluate to produce a single result. An assignment binds a name to that result. The two always appear together: you evaluate an expression on the right of =, and the name on the left receives the value.

Compound Assignment Operators

Compound assignment operators combine an arithmetic operation with assignment in one step: +=, -=, *=, /=, //=, %=, and **=. They are shorthand for the pattern variable = variable OP value and are cleaner when updating a running value. They do not return a value themselves and cannot be used inside a larger expression.

Example 1 — Introduction

Compound assignment operators in a session-based financial accumulator.

```
total_revenue   = 0.0
total_refunds   = 0.0
transaction_fee = 0.015  # 1.5% per transaction

transactions = [
    ("sale",    849.00),
    ("sale",  1_240.50),
    ("refund",  200.00),
    ("sale",    390.75),
    ("refund",   85.00),
]

for type_, amount in transactions:
    fee = amount * transaction_fee
    if type_ == "sale":
        total_revenue += amount - fee
    else:
        total_refunds += amount
```

```
net = total_revenue - total_refunds
print(f"Revenue (net of fees): ${total_revenue:,.2f}")
print(f"Refunds:               ${total_refunds:,.2f}")
print(f"Net:                   ${net:,.2f}")
```

Multi-line and Complex Expressions

Python allows expressions to span multiple lines when they are enclosed in parentheses. Use this to break long formulas into labelled sub-expressions that read like the mathematical specification they implement. Each intermediate value can be assigned a meaningful name, making the calculation auditable and the intent obvious.

Example 2 — Variation

A multi-line expression for actuarial net present value, broken into labelled components.

```
import math

DISCOUNT_RATE = 0.07
INFLATION     = 0.025

cash_flows = [120_000, 135_000, 148_000, 162_000, 180_000]

# Real discount rate (Fisher equation)
real_rate = (1 + DISCOUNT_RATE) / (1 + INFLATION) - 1

npv = sum(
    cf / (1 + real_rate) ** (year + 1)
    for year, cf in enumerate(cash_flows)
)

print(f"Real discount rate: {real_rate:.4f}")
print(f"NPV (real terms):   ${npv:,.0f}")
```

5.4 Using math Library Functions

Python's math module provides mathematical functions and constants that go beyond what the arithmetic operators cover. It includes logarithms, trigonometric functions, rounding variants, combinatorial functions, and the constants pi and e. Import it with import math and access its members with the math. prefix.

A selection of the most commonly used functions: math.sqrt(x) returns the square root as float; math.pow(x, y) returns x ** y as float (unlike the ** operator which preserves int); math.log(x) is the natural log, math.log(x, base) for any base, math.log10(x) for base 10; math.ceil(x) rounds up to the nearest integer; math.floor(x) rounds down; math.trunc(x) truncates toward zero; math.factorial(n) returns n! as an exact integer; math.degrees(r) and math.radians(d) convert between angle units; math.pi and math.e are the standard constants.

Logarithms and Roots

Example 1 — Introduction

Logarithms and roots in financial and signal-processing contexts.

```
import math

# Compound annual growth rate (CAGR)
# CAGR = (end / start) ** (1/years) - 1
start_value = 1_500_000
end_value   = 2_840_000
years       = 6
cagr = (end_value / start_value) ** (1 / years) - 1
print(f"CAGR: {cagr:.2%}")

# Doubling time via Rule of 72 vs exact logarithm
rate = 0.065
rule_of_72    = 72 / (rate * 100)
exact_years   = math.log(2) / math.log(1 + rate)
print(f"Rule of 72:  {rule_of_72:.2f} years")
print(f"Exact:       {exact_years:.2f} years")

# Signal amplitude in decibels: dB = 20 * log10(V2/V1)
v_ref, v_sig = 1.0, 3.162
decibels = 20 * math.log10(v_sig / v_ref)
print(f"Signal level: {decibels:.1f} dB")
```

Trigonometry and Combinatorics

Example 2 — Variation

Trigonometric functions for a surveying calculation; factorial for a probability model.

```
import math

# Surveying: horizontal distance from slope distance and grade angle
slope_distance_m = 147.3
grade_degrees    = 8.5
grade_radians    = math.radians(grade_degrees)

horizontal = slope_distance_m * math.cos(grade_radians)
vertical   = slope_distance_m * math.sin(grade_radians)
print(f"Horizontal: {horizontal:.2f} m")
print(f"Vertical:   {vertical:.2f} m")

# Probability: chance of k successes in n trials (binomial coefficient)
# C(n, k) = n! / (k! * (n-k)!)
n, k = 20, 3
combinations = math.factorial(n) // (math.factorial(k) * math.factorial(n -
k))
print(f"C({n},{k}) = {combinations:,}")
```

5.5 Rounding and Truncating

Controlling numeric precision is a practical necessity in almost every domain. Currency calculations must produce exactly two decimal places. Scientific measurements carry meaningful significant figures. Integer counts cannot have fractional parts. Python provides four distinct mechanisms, each with different behaviour, and choosing the right one matters.

round(), math.ceil(), math.floor(), and math.trunc()

round(x, n) rounds x to n decimal places using banker's rounding (rounds half to even). round(x) with no second argument returns an int. math.ceil(x) rounds up to the nearest integer regardless of the fractional part. math.floor(x) rounds down. math.trunc(x) truncates toward zero (positive numbers behave like floor, negative numbers like ceil).

Example 1 — Introduction

All four rounding functions compared on the same value, then applied in a billing context.

```
import math

value = 7.65
print(f"round(7.65, 1):    {round(value, 1)}")  # 7.6 (banker's)
print(f"math.ceil(7.65):   {math.ceil(value)}")  # 8
print(f"math.floor(7.65):  {math.floor(value)}") # 7
print(f"math.trunc(7.65):  {math.trunc(value)}") # 7

# Billing: always round up to next cent (ceil to 2 d.p.)
def round_up_cents(amount: float) -> float:
    return math.ceil(amount * 100) / 100

charges = [14.321, 8.001, 22.005, 5.999]
for c in charges:
    print(f"  {c:.3f}  →  ${round_up_cents(c):.2f}")
```

Banker's Rounding

Python's round() uses banker's rounding (also called round-half-to-even), which rounds .5 to the nearest even digit rather than always up. This reduces systematic bias in large datasets: round(0.5) is 0, round(1.5) is 2, round(2.5) is 2, round(3.5) is 4. For financial calculations where always-round-up-on-halfway is required, use the Decimal module with explicit rounding modes.

Example 2 — Variation

Banker's rounding vs conventional rounding, and Decimal with ROUND_HALF_UP for financial precision.

```
from decimal import Decimal, ROUND_HALF_UP

# Banker's rounding (Python default)
for x in [0.5, 1.5, 2.5, 3.5, 4.5]:
    print(f"  round({x}) = {round(x)}")

# ROUND_HALF_UP for financial/statutory requirements
TWO_PLACES = Decimal("0.01")
```

```
values = ["14.325", "8.315", "22.165", "5.995"]
for v in values:
    d      = Decimal(v)
    banker = float(round(d, 2))
    halfup = float(d.quantize(TWO_PLACES, rounding=ROUND_HALF_UP))
    print(f"  {v}  banker={banker:.2f}  half_up={halfup:.2f}")
```

5.6 Practical Arithmetic Problems

The operators, functions, and rounding tools from earlier in this chapter combine to solve real computational problems. The five examples below demonstrate progressively complex applied arithmetic across finance, engineering, and logistics domains.

Compound Interest and Loan Amortisation

Example 1 — Introduction

Compound interest formula and full loan amortisation for the first three payments.

```
import math

# Compound interest: future value
principal    = 10_000
annual_rate  = 0.048
compounds    = 12          # monthly
years        = 5

future_value = principal * (1 + annual_rate / compounds) ** (compounds *
years)
print(f"Future value: ${future_value:,.2f}")

# Monthly repayment (standard amortisation formula)
loan         = 320_000
monthly_rate = 0.0589 / 12
n_months     = 25 * 12

payment = loan * monthly_rate * (1 + monthly_rate) ** n_months \
           / ((1 + monthly_rate) ** n_months - 1)

print(f"\nMonthly payment: ${payment:,.2f}")
balance = loan
for month in range(1, 4):
    interest   = balance * monthly_rate
    principal_ = payment - interest
    balance   -= principal_
    print(f"  Month {month}: interest ${interest:.2f}, principal
${principal_:.2f}, balance ${balance:,.2f}")
```

Quadratic Equation Solver

Example 2 — Variation

A general quadratic solver used to find break-even units in a cost/revenue model.

```
import math
```

```
def solve_quadratic(a: float, b: float, c: float):
    """
    Solve ax^2 + bx + c = 0.
    Returns a tuple of real roots, or None if no real roots exist.
    """
    discriminant = b ** 2 - 4 * a * c
    if discriminant > 0:
        r1 = (-b + math.sqrt(discriminant)) / (2 * a)
        r2 = (-b - math.sqrt(discriminant)) / (2 * a)
        return (r1, r2)
    elif discriminant == 0:
        return (-b / (2 * a),)
    return None

# Break-even: revenue = cost
# R(x) = 85x      (revenue per unit)
# C(x) = 0.2x^2 + 30x + 15000   (total cost with quadratic overhead)
# 85x = 0.2x^2 + 30x + 15000
# 0 = 0.2x^2 - 55x + 15000
roots = solve_quadratic(0.2, -55, 15_000)
if roots:
    for r in roots:
        print(f"Break-even at {r:.0f} units")
```

Unit Conversion and Engineering Calculations

Example 3 — Real-world application

A multi-unit pipeline flow calculator converting between unit systems.

```
import math

# Pipeline flow: velocity and volume flow rate
# Given: pipe diameter in mm, flow speed in m/s

diameter_mm  = 150.0
flow_speed   = 2.4     # m/s

radius_m     = (diameter_mm / 1000) / 2
area_m2      = math.pi * radius_m ** 2
flow_m3s     = area_m2 * flow_speed

# Convert to common engineering units
flow_litres_s   = flow_m3s * 1_000
flow_litres_min = flow_litres_s * 60
flow_m3_hr      = flow_m3s * 3_600

print(f"Pipe diameter:   {diameter_mm:.0f} mm")
print(f"Cross-section:   {area_m2*10_000:.2f} cm²")
print(f"Flow rate:       {flow_litres_s:.2f} L/s")
print(f"               = {flow_litres_min:.1f} L/min")
print(f"               = {flow_m3_hr:.2f} m³/hr")
```

5.7 Avoiding Common Arithmetic Mistakes

Several classes of arithmetic error appear repeatedly in Python code. Understanding them in advance — and knowing the correct pattern for each — prevents bugs that can be difficult to diagnose after the fact.

Floating-Point Precision Errors

Floating-point numbers are stored in IEEE 754 binary format, which cannot represent every decimal fraction exactly. The canonical example is 0.1 + 0.2, which does not equal 0.3 in binary floating-point. This is not a bug in Python — it is an unavoidable consequence of representing base-10 decimals in base-2. The issue matters most in financial calculations, where apparently small errors accumulate across thousands of transactions.

Example 1 — Introduction

Floating-point accumulation error in a running total, and two correct approaches.

```
from decimal import Decimal

# The problem: float representation error accumulates
prices_float = [0.10, 0.20, 0.30] * 1_000
total_float  = sum(prices_float)
print(f"Float total:   {total_float:.6f}") # not exactly 600.0

# Solution 1: round() for display (masks the error, does not fix it)
print(f"Rounded:       {round(total_float, 2)}")

# Solution 2: Decimal for exact arithmetic (correct approach for finance)
prices_dec  = [Decimal("0.10"), Decimal("0.20"), Decimal("0.30")] * 1_000
total_dec   = sum(prices_dec)
print(f"Decimal total: {total_dec}")       # exactly 600.00
```

Integer Division and Type Surprises

/ always returns float in Python 3. This means 10 / 5 is 2.0, not 2. If you need an integer result, use //. More subtly: int(10 / 5) works for small numbers but can fail for large values because the float intermediate may lose precision. Prefer 10 // 5 when you need integer division.

Example 2 — Variation

Division type behaviour and the preferred integer division pattern.

```
# / always returns float
print(10 / 5)      # 2.0  (float)
print(10 // 5)     # 2    (int)
print(type(10 / 5))  # <class 'float'>
print(type(10 // 5)) # <class 'int'>

# Subtle failure with large numbers: int(large_num / divisor) may be wrong
# because large_num/divisor loses integer precision as a float
large = 10 ** 18 + 1
divisor = 3
print(large // divisor)         # exact integer division
print(int(large / divisor))     # may differ: float lost precision
```

```
# Percentage calculation: remember to use float division
correct_count = 87
total         = 120
pct = correct_count / total * 100   # / gives float: 72.5
print(f"Score: {pct:.1f}%")
```

Modulus with Negative Numbers

Python's % follows the sign of the divisor, not the dividend. This is mathematically consistent with floor division but differs from C, Java, and JavaScript where % follows the dividend's sign. Always test modulus expressions with negative inputs if they may receive them.

Example 3 — Real-world application

Modulus sign behaviour and a robust hour-normalisation function that handles negative inputs.

```
# Python: sign follows the divisor
print(-7 % 3)   #  2  (positive divisor)
print( 7 % -3)  # -2  (negative divisor)

# C/Java/JS:  -7 % 3 == -1 -- different!

# Normalising a clock offset to [0, 24) correctly handles negatives
def normalise_hour(hour: int) -> int:
    """Return hour modulo 24, always in range [0, 23]."""
    return hour % 24   # Python modulus handles negatives correctly

offsets = [-1, 0, 23, 24, 25, -13]
for h in offsets:
    print(f"  hour {h:>3}  →  {normalise_hour(h):>2}:00")
```

Mixing Types Unexpectedly

Python promotes int to float in mixed arithmetic, which is usually correct but occasionally surprising. A more dangerous pattern is accidentally computing integer division when float division was intended because both operands happen to be int. And int(float_value) truncates toward zero — it does not round. Use round() when you need rounded-to-nearest; use int() only when truncation is explicitly required.

```
# Accidental integer arithmetic in a percentage formula
# If both values are int, // gives int -- fine here, but verify intent
rate_bps  = 325     # basis points (int)
rate_pct  = rate_bps / 100    # 3.25% (float -- / always gives float)

# int() truncates, does not round
print(int(3.9))    # 3, not 4
print(round(3.9))  # 4
print(int(-3.9))   # -3 (truncates toward zero, not floor)
print(math.floor(-3.9))  # -4 (actual floor)
```

Chapter 5 Summary

- Python provides seven arithmetic operators: + (add), - (subtract/negate), * (multiply), / (true division — always returns float), // (floor division), % (modulus — sign follows divisor), and ** (exponentiation — right-associative).
- Operator precedence from highest to lowest: () → ** → unary +/- → *, /, //, % → +, -. When in doubt, add parentheses — they are documentation as much as syntax.
- Compound assignment operators (+=, -=, *=, /=, //=, %=, **=) update a variable in place. Use them to accumulate running totals and update state in loops.
- The math module provides: sqrt(), log(), log10(), log2(), pow(), sin(), cos(), tan(), ceil(), floor(), trunc(), factorial(), radians(), degrees(), and the constants math.pi and math.e.
- round(x, n) uses banker's rounding (round-half-to-even). math.ceil() always rounds up; math.floor() always rounds down; math.trunc() truncates toward zero. int() also truncates toward zero — it does not round.
- Float arithmetic cannot represent all decimals exactly. $0.1 + 0.2 \neq 0.3$ in binary floating-point. For financial calculations requiring exact decimal arithmetic, use the decimal.Decimal type with explicit rounding modes.
- / always returns float in Python 3 even for two int operands. Use // for integer floor division. Use % carefully with negative numbers: result sign follows the divisor, not the dividend.

Guided Exercises

Complete the following exercises in VS Code, IDLE, or your preferred Python environment. Each exercise targets a specific concept from this chapter.

1. Write a stock reconciliation script. Define opening_stock, units_received, units_dispatched, and units_damaged as variables. Calculate closing_stock and variance against an expected count. Use // and % to determine how many full pallets (24 units each) close stock represents and how many loose units remain. Print a formatted summary using f-strings with alignment specifiers.

2. Implement a loan amortisation calculator. Accept principal, annual interest rate, and loan term in years as inputs. Compute the monthly payment using the standard formula. Print the first six months of the amortisation schedule showing: month number, interest paid, principal repaid, and remaining balance. Format all currency values to two decimal places with thousands separators.

3. Write a compound interest comparison function that accepts a principal amount, annual rate, and number of years, and prints a table comparing simple interest against monthly, quarterly, and annual compounding. Use math.pow() for one column and ** for another to confirm they produce identical results. Apply appropriate rounding to each figure.

4. Demonstrate floating-point precision issues by summing a list of 1,000 identical float values (e.g., 0.1) and comparing the result against the expected total using ==. Then implement the correct solution using decimal.Decimal with string literals, and show that the Decimal sum is exact. Explain in a comment why the float version fails.

5. Using the math module, write a pipeline flow calculator. Accept pipe diameter (mm) and flow velocity (m/s). Compute and print: cross-sectional area (cm^2), volumetric flow in L/s, L/min, and m^3/hr. Also compute the Reynolds number given a kinematic viscosity input, and classify the flow as laminar (Re < 2300), transitional (2300–4000), or turbulent (> 4000).

Challenge Tasks

The following tasks are open-ended. No hints are provided. They require you to design a solution, not just apply a pattern.

1. Build a complete financial modelling tool that accepts: an initial investment, a sequence of annual cash flows (up to 10 years), and a discount rate. Compute and display: the net present value (NPV), the internal rate of return (IRR, found by bisection search between -50% and +500%), the payback period (years to recover initial investment), and a year-by-year table showing undiscounted and discounted cumulative cash flow. Use Decimal for all monetary arithmetic and ROUND_HALF_UP for all displayed figures.

2. Write a rounding comparison tool. For a configurable list of monetary amounts (include values with exactly .5 cents, .005 cents, and values near round numbers), compute and display the result of six different approaches: Python float round(), math.ceil() to 2 d.p., math.floor() to 2 d.p., Decimal ROUND_HALF_UP, Decimal ROUND_HALF_EVEN, and Decimal ROUND_UP. Show total discrepancies between the most and least aggressive rounding strategies across the full list. Include at least 12 test values chosen to expose differences between methods.

3. Design and implement a generalised unit conversion library. It should support at least three categories: length (mm, cm, m, km, inches, feet, miles, nautical miles), mass (mg, g, kg, tonnes, oz, lb), and data size (bits, bytes, KB, MB, GB, TB). Write a convert(value, from_unit, to_unit) function that performs any conversion within a category, raising ValueError for cross-category conversions or unknown units. Build a command-line interface that accepts value, from_unit, to_unit as inputs, validates them, performs the conversion, and prints the result with appropriate precision. Test with at least fifteen conversion pairs including several that chain through a base unit.

CHAPTER 6 Comparison and Logical Operators

Think, Code, Create – A Python Programming Journey

Every program that does something useful makes decisions. It checks whether a value is within range, whether a user has permission, whether a transaction should be approved, whether a sensor reading indicates a fault. These decisions are expressed in Python using comparison operators — which evaluate a relationship between two values and produce a boolean result — and logical operators, which combine boolean values into compound conditions.

This chapter covers all six comparison operators, the three logical operators, truth tables, chaining and short-circuit evaluation, and how these tools compose into the multi-condition logic that drives real systems. Section 6.5 through 6.7 demonstrate the same concepts in progressively complex applied contexts — access control, data filtering, and automated control systems. By the end of this chapter you will be able to write and reason about compound boolean expressions with confidence.

6.1 Boolean Values and Comparisons

A boolean is a value that is either True or False. In Python, bool is a built-in type and a subclass of int — True equals 1 and False equals 0. Every comparison expression evaluates to a bool, and bools are what control the flow of if statements, while loops, and any construct that makes a decision.

All six comparison operators compare two values and return a boolean: == (equal to), != (not equal to), < (less than), > (greater than), <= (less than or equal to), >= (greater than or equal to). The operands can be numbers, strings, dates, or any other type that supports the given comparison.

Comparison Operators in Practice

Example 1 — Introduction

Comparison operators applied to a server health monitoring context.

```
response_time_ms  = 847
error_rate_pct    = 2.3
active_connections = 1_204

RESPONSE_WARN  = 500
RESPONSE_CRIT  = 1_000
ERROR_THRESHOLD = 5.0
MAX_CONNECTIONS = 2_000

print(response_time_ms > RESPONSE_WARN)     # True  -- degraded
print(response_time_ms >= RESPONSE_CRIT)    # False -- not critical yet
print(error_rate_pct == ERROR_THRESHOLD)    # False
print(error_rate_pct < ERROR_THRESHOLD)     # True  -- within limits
```

```
print(active_connections != MAX_CONNECTIONS) # True
print(active_connections <= MAX_CONNECTIONS) # True
```

Chained Comparisons

Python allows comparison operators to be chained: a < b < c is evaluated as (a < b) and (b < c), checking that b lies between a and c. This is more readable than the equivalent compound expression and evaluates b only once. Chaining works with any mix of comparison operators.

Example 2 — Variation

Chained comparisons checking sensor readings against valid operating ranges.

```
TEMP_MIN, TEMP_MAX       = 15.0, 85.0   # Celsius
VOLTAGE_MIN, VOLTAGE_MAX = 11.5, 13.0  # Volts
PRESSURE_MIN, PRESSURE_MAX = 0.8, 1.2  # bar

temperature = 72.4
voltage     = 12.1
pressure    = 0.95

# Chained comparisons -- cleaner than: temp >= MIN and temp <= MAX
temp_ok     = TEMP_MIN     <= temperature <= TEMP_MAX
voltage_ok  = VOLTAGE_MIN  <= voltage     <= VOLTAGE_MAX
pressure_ok = PRESSURE_MIN <= pressure    <= PRESSURE_MAX

print(f"Temperature: {temperature}C  in range: {temp_ok}")
print(f"Voltage:     {voltage}V      in range: {voltage_ok}")
print(f"Pressure:    {pressure} bar  in range: {pressure_ok}")
```

Comparing Strings and Other Types

Strings are compared lexicographically (character by character, using Unicode code points). This means "apple" < "banana" is True, and "Z" < "a" is True because uppercase letters have lower code points than lowercase. For case-insensitive comparison, normalise both strings first with .lower() or .casefold(). Comparing values of incompatible types — such as an int and a list — raises a TypeError.

Example 3 — Real-world application

String comparison for case-insensitive status checking and sorting product codes.

```
status_raw = "  Active  "

# Always normalise before comparing string input
is_active  = status_raw.strip().lower() == "active"
print(f"Is active: {is_active}")

# String ordering: useful for sorting and range checks
product_codes = ["C-200", "A-100", "B-150", "A-050"]
product_codes.sort()
print("Sorted:", product_codes)

# Range check on string codes
code = "B-075"
```

```
in_range = "A-000" <= code <= "B-999"
print(f"{code} in A-B range: {in_range}")
```

6.2 Logical Operators: and, or, not

The three logical operators combine boolean expressions into compound conditions. and returns True only when both operands are True. or returns True when at least one operand is True. not inverts its single operand. These operators work on any value, not just explicit True/False literals, because Python evaluates any value in a boolean context: zero, empty containers, and None are falsy; everything else is truthy.

and, or, not

Example 1 — Introduction

Logical operators in a financial transaction approval system.

```
account_balance  = 4_850.00
credit_limit     = 2_000.00
is_verified      = True
is_suspended     = False
transaction_amt  = 1_200.00

# and: ALL conditions must be True
can_approve = (
    is_verified
    and not is_suspended
    and (account_balance + credit_limit) >= transaction_amt
)
print(f"Approved: {can_approve}")

# or: ANY condition triggers a review flag
needs_review = (
    transaction_amt > 5_000
    or is_suspended
    or not is_verified
)
print(f"Needs review: {needs_review}")
```

Short-Circuit Evaluation

Python evaluates logical expressions lazily: it stops as soon as the result is determined. For and, if the left operand is falsy, the right operand is never evaluated — the result is already False. For or, if the left operand is truthy, the right operand is never evaluated — the result is already True. This is not merely an optimisation; it is a language guarantee that can be used deliberately to prevent errors.

Example 2 — Variation

Short-circuit evaluation preventing a ZeroDivisionError and a KeyError.

```
# Safe division: right side not evaluated when divisor is zero
total_calls   = 0
failed_calls  = 0
```

```
error_rate     = total_calls > 0 and failed_calls / total_calls
print(f"Error rate: {error_rate}")    # False, not a ZeroDivisionError

# Safe key lookup: dict.get() preferred, but short-circuit also works
config = {"timeout": 30}

has_retry      = "retry" in config and config["retry"] > 0
timeout        = "timeout" in config and config["timeout"]
print(f"Has retry: {has_retry}")      # False (key absent, right not eval)
print(f"Timeout:   {timeout}")        # 30 (truthy integer)
```

Truthiness and Falsiness

Every Python value has a boolean interpretation. Falsy values are: False, None, 0, 0.0, 0j, empty string "", empty list [], empty tuple (), empty dict {}, empty set set(). Everything else is truthy. This means you can write if items: instead of if len(items) > 0:, or if username: instead of if username != "":. Using truthiness directly makes code more idiomatic and readable.

Example 3 — Real-world application

Using truthiness to guard processing steps — cleaner than explicit length or None checks.

```
def process_batch(records: list, config: dict) -> dict:
    """Process a batch of records using config settings."""
    # Truthy checks replace explicit len() and != None comparisons
    if not records:
        return {"status": "skipped", "reason": "empty batch"}

    timeout   = config.get("timeout") or 30    # fallback if missing/zero
    max_retry = config.get("max_retry") or 3

    processed = []
    for record in records:
        if record.get("active") and record.get("value"):
            processed.append(record)

    return {"status": "ok", "count": len(processed), "timeout": timeout}

result = process_batch([], {"timeout": 0})
print(result)
result = process_batch([{"active": True, "value": 42}], {"timeout": 60})
print(result)
```

6.3 Truth Tables

A truth table enumerates every possible combination of input boolean values and shows the corresponding output for a logical expression. Truth tables are a compact and definitive way to verify that a compound boolean expression behaves correctly — particularly useful when debugging conditions that produce unexpected results or when designing business rules that must be formally correct.

AND, OR, and NOT

Example 1 — Introduction

The three fundamental truth tables, verified by Python code.

```
# AND truth table
# A       B       A and B
# True    True    True
# True    False   False
# False   True    False
# False   False   False

# OR truth table
# A       B       A or B
# True    True    True
# True    False   True
# False   True    True
# False   False   False

# NOT truth table
# A       not A
# True    False
# False   True

# Verify programmatically
for A in [True, False]:
    for B in [True, False]:
        print(f"A={str(A):<5} B={str(B):<5} AND={str(A and B):<5} OR={str(A or
B):<5}")
```

Compound Expressions

Truth tables extend naturally to compound expressions. To verify any multi-operator expression, iterate over every combination of its boolean inputs and print the result. This is a practical debugging technique: if the table for your condition does not match your intention, you can find the exact combination of inputs that produces the wrong result.

Example 2 — Variation

Truth table for a loan approval rule: approved = is_employed and (credit_ok or has_collateral).

```
print(f"{"employed":<10} {"credit_ok":<10} {"collateral":<12} approved")
print("-" * 46)
for is_employed in [True, False]:
    for credit_ok in [True, False]:
        for has_collateral in [True, False]:
            approved = is_employed and (credit_ok or has_collateral)
            print(
                f"{str(is_employed):<10}"
                f"{str(credit_ok):<10}"
                f"{str(has_collateral):<12}"
                f"{approved}"
            )
```

Reading the table confirms: an applicant is approved only when employed AND (has a good credit score OR provides collateral). An unemployed applicant is always rejected regardless of the other factors. This matches the business rule intent and any row that produces an unexpected result reveals a logical error in the expression.

De Morgan's Laws

De Morgan's laws provide algebraically equivalent transformations for boolean expressions involving not with and and or. They are useful when simplifying conditions or reasoning about what a negated condition actually means:

not (A and B) is equivalent to (not A) or (not B)

not (A or B) is equivalent to (not A) and (not B)

Example 3 — Real-world application

De Morgan's laws: refactoring a rejection condition into its approval equivalent.

```
# Original rule expressed as a rejection condition:
# "Reject if NOT (verified AND active)"
def is_rejected_original(is_verified, is_active):
    return not (is_verified and is_active)

# De Morgan transform: not(A and B) == (not A) or (not B)
# "Reject if unverified OR inactive"
def is_rejected_demorgan(is_verified, is_active):
    return (not is_verified) or (not is_active)

# Both are logically identical
for v in [True, False]:
    for a in [True, False]:
        orig = is_rejected_original(v, a)
        dmgn = is_rejected_demorgan(v, a)
        assert orig == dmgn, f"Mismatch at v={v}, a={a}"
        print(f"verified={str(v):<5} active={str(a):<5} rejected={orig}")
```

6.4 Nested and Compound Conditions

Compound conditions combine multiple comparisons in a single expression using logical operators. Nested conditions place one conditional inside another to express dependent logic: the inner condition is only evaluated when the outer condition has already been satisfied. Both patterns appear constantly in real programs; understanding when to use each — and how to keep them readable — is an important skill.

Compound Conditions

Compound conditions use and, or, and not to combine comparisons. Keep compound conditions readable by: using parentheses to make grouping explicit, breaking very long conditions across multiple lines inside parentheses, and assigning named boolean variables when individual sub-conditions have distinct meaning.

Example 1 — Introduction

A compound approval condition for a property loan, written for readability.

```
INCOME_MIN    = 80_000
CREDIT_MIN    = 680
LVR_MAX       = 0.90     # loan-to-value ratio
```

```
DTI_MAX      = 0.43     # debt-to-income ratio

annual_income = 105_000
credit_score  = 715
loan_amount   = 550_000
property_val  = 680_000
monthly_debt  = 2_400

# Named booleans make the compound condition auditable
income_ok = annual_income >= INCOME_MIN
credit_ok = credit_score  >= CREDIT_MIN
lvr_ok    = (loan_amount / property_val) <= LVR_MAX
dti_ok    = (monthly_debt / (annual_income / 12)) <= DTI_MAX

approved  = income_ok and credit_ok and lvr_ok and dti_ok

print(f"Income ok:  {income_ok}")
print(f"Credit ok:  {credit_ok}")
print(f"LVR ok:     {lvr_ok}")
print(f"DTI ok:     {dti_ok}")
print(f"Approved:   {approved}")
```

Nested Conditions

Nested conditions check a secondary condition only after an outer condition has passed. They are appropriate when the inner condition only makes sense in the context of the outer one. Excessive nesting — more than two or three levels — is usually a signal to extract logic into named functions. A flat sequence of if/elif/else is often easier to read than deeply nested ifs.

Example 2 — Variation

Nested conditions in a user permission system: outer checks authentication, inner checks role.

```
def check_access(user: dict, resource: str) -> str:
    if not user.get("is_authenticated"):
        return "Access denied: not authenticated"

    if user.get("is_suspended"):
        return "Access denied: account suspended"

    role = user.get("role", "viewer")
    if resource == "admin_panel":
        if role == "admin":
            return "Access granted: admin panel"
        else:
            return f"Access denied: {role} cannot access admin panel"
    elif resource == "reports":
        if role in ("admin", "analyst", "manager"):
            return "Access granted: reports"
        else:
            return f"Access denied: {role} cannot access reports"
    return "Access granted"

users = [
    {"is_authenticated": True,  "role": "admin",   "is_suspended": False},
    {"is_authenticated": True,  "role": "viewer",  "is_suspended": False},
```

```
    {"is_authenticated": False, "role": "admin",   "is_suspended": False},
]
for u in users:
    print(check_access(u, "admin_panel"))
```

Flattening Nested Conditions with Early Return

Deeply nested conditions can often be restructured using early return (also called guard clauses): check each failure condition first and return immediately, leaving the successful path as the main body of the function with minimal indentation. This pattern is generally more readable than deeply nested else blocks.

Example 3 — Real-world application

Refactoring nested conditions into early-return guard clauses in a data validation function.

```
# Deeply nested (hard to follow)
def validate_nested(order):
    if order:
        if order.get("customer_id"):
            if order.get("items"):
                if order.get("total", 0) > 0:
                    return True
    return False

# Flat with guard clauses (easier to read)
def validate_flat(order):",
    if not order:                        return False
    if not order.get("customer_id"):     return False
    if not order.get("items"):           return False
    if not order.get("total", 0) > 0:   return False
    return True

test_orders = [
    None,
    {},
    {"customer_id": "C001"},
    {"customer_id": "C001", "items": ["A"], "total": 49.95},
]
for o in test_orders:
    print(f"nested={validate_nested(o)}  flat={validate_flat(o)}")
```

6.5 Real-World Boolean Problems

Boolean logic is the mechanism that turns data into decisions. The examples below show how comparison and logical operators compose to solve authentication, eligibility checking, and data filtering problems that appear in production systems.

Multi-Factor Authentication Check

Example 1 — Introduction

An authentication gate that checks credentials, account status, and session validity together.

```
from datetime import datetime
```

```python
def authenticate(user: dict, provided_token: str) -> tuple[bool, str]:
    """
    Returns (is_authenticated, reason).
    All conditions must pass; the first failure short-circuits evaluation.
    """
    if not user.get("is_active"):
        return False, "Account is deactivated"

    if user.get("failed_attempts", 0) >= 5:
        return False, "Account locked after too many failures"

    if provided_token != user.get("token"):
        return False, "Invalid token"

    expiry = user.get("token_expiry")
    if expiry and datetime.now() > expiry:
        return False, "Token expired"

    return True, "OK"

user = {
    "is_active": True,
    "failed_attempts": 0,
    "token": "abc123",
    "token_expiry": None,
}
ok, reason = authenticate(user, "abc123")
print(f"Auth: {ok} ({reason})")
```

Data Filtering with Boolean Expressions

Example 2 — Variation

Filtering a product catalogue by multiple independent criteria using list comprehension.

```python
products = [
    {"sku": "A-100", "price": 12.50, "stock": 45,  "active": True,
"category": "hardware"},
    {"sku": "B-200", "price": 89.99, "stock": 0,   "active": True,
"category": "software"},
    {"sku": "C-300", "price": 5.00,  "stock": 200, "active": False,
"category": "hardware"},
    {"sku": "D-400", "price": 34.95, "stock": 12,  "active": True,
"category": "hardware"},
    {"sku": "E-500", "price": 150.00,"stock": 8,   "active": True,
"category": "software"},
]

# Filter: active hardware products in stock under $50
available = [
    p for p in products
    if p["active"]
    and p["stock"] > 0
    and p["category"] == "hardware"
    and p["price"] < 50
]
```

```
print(f"Matching products: {len(available)}")
for p in available:
    print(f"  {p['sku']}: ${p['price']:.2f} ({p['stock']} in stock)")
```

Eligibility Scoring

Example 3 — Real-world application

A credit pre-qualification system that scores each criterion separately and reports all failures.

```
def credit_prequalify(applicant: dict) -> dict:
    """
    Check each criterion independently and return a full result.
    Unlike short-circuit evaluation, this reports ALL failures.
    """
    checks = {
        "income":      applicant.get("annual_income", 0)    >= 60_000,
        "credit":      applicant.get("credit_score", 0)     >= 650,
        "employment":  applicant.get("employment_months", 0) >= 12,
        "dti":         applicant.get("debt_to_income", 1)  <= 0.43,
        "no_defaults": not applicant.get("has_defaults", False),
    }
    approved = all(checks.values())
    failures = [k for k, v in checks.items() if not v]
    return {"approved": approved, "checks": checks, "failures": failures}

applicant = {
    "annual_income": 72_000, "credit_score": 620,
    "employment_months": 18, "debt_to_income": 0.38,
    "has_defaults": False,
}
result = credit_prequalify(applicant)
print(f"Approved: {result['approved']}")
if result["failures"]:
    print(f"Failed criteria: {result['failures']}")
```

6.6 Logical Problem: Automated Alert System

Alert systems are a canonical application of compound boolean logic. They must fire exactly when conditions warrant — not when they don't — and must handle multiple independent signals at different severity levels. Getting the boolean logic right is critical: a false negative means a real problem goes undetected; a false positive means engineers respond to noise.

Single-Service Health Check

Example 1 — Introduction

A service health checker that classifies severity from multiple metrics using compound boolean logic.

```
def classify_health(metrics: dict) -> str:
    """
    Classify a service as healthy, degraded, or critical.
    Criteria are checked in order of severity.
    """
```

```
    cpu     = metrics.get("cpu_pct",    0)
    mem     = metrics.get("mem_pct",    0)
    errors = metrics.get("error_pct", 0)
    latency = metrics.get("latency_ms", 0)

    # Critical: any one of these alone is a page-worthy incident
    is_critical = (
        cpu >= 95
        or mem >= 95
        or errors >= 10
        or latency >= 5_000
    )
    if is_critical:
        return "CRITICAL"

    # Degraded: two or more metrics elevated but not critical
    elevated = sum([cpu >= 75, mem >= 80, errors >= 2, latency >= 1_000])
    if elevated >= 2:
        return "DEGRADED"

    return "HEALTHY"

services = {
    "api-gateway":  {"cpu_pct": 62, "mem_pct": 55, "error_pct": 0.2,
"latency_ms": 180},
    "auth-service": {"cpu_pct": 88, "mem_pct": 91, "error_pct": 1.5,
"latency_ms": 950},
    "db-primary":   {"cpu_pct": 97, "mem_pct": 70, "error_pct": 0.0,
"latency_ms": 12},
}

for name, m in services.items():
    print(f"  [{classify_health(m):>8}]  {name}")
```

Multi-Service Escalation Logic

Example 2 — Variation

Escalation policy: page on-call only when at least one critical service, or two or more degraded ones.

```
def should_page_on_call(service_statuses: list[str]) -> tuple[bool, str]:
    """
    Determine whether to trigger an on-call page.
    Returns (should_page, reason).
    """
    critical_count = service_statuses.count("CRITICAL")
    degraded_count = service_statuses.count("DEGRADED")

    if critical_count >= 1:
        return True, f"{critical_count} critical service(s)"

    if degraded_count >= 2:
        return True, f"{degraded_count} services degraded simultaneously"

    return False, "All services within normal parameters"
```

```
statuses = ["HEALTHY", "DEGRADED", "CRITICAL"]
page, reason = should_page_on_call(statuses)
if page:
    print(f"ALERT: Paging on-call engineer -- {reason}")
else:
    print(f"OK: {reason}")
```

6.7 Combining Boolean Expressions in Complex Systems

Real systems combine boolean conditions across multiple dimensions simultaneously. The examples in this section show how the same logical building blocks — and, or, not, chained comparisons, truthiness checks — compose into the decision logic of a smart building controller and an inventory management system.

Smart Building Controller

Example 1 — Introduction

A building management system that controls HVAC and lighting based on occupancy, time, and sensor data.

```
def control_hvac(zone: dict) -> str:
    """Determine HVAC action for a zone based on current conditions."""
    is_occupied    = zone.get("occupancy_count", 0) > 0
    temperature    = zone.get("temperature_c", 20)
    is_after_hours = zone.get("is_after_hours", False)
    override       = zone.get("manual_override")

    if override:
        return f"Manual override: {override}"

    if not is_occupied and is_after_hours:
        return "Setback mode: 18°C economy target"

    if is_occupied and temperature < 19.0:
        return "Heating: zone below comfort range"

    if is_occupied and temperature > 24.0:
        return "Cooling: zone above comfort range"

    return "Maintain: temperature within comfort range"

zones = [
    {"name": "Level 3 Open Plan", "occupancy_count": 24,
     "temperature_c": 26.2, "is_after_hours": False},
    {"name": "Server Room",       "occupancy_count": 0,
     "temperature_c": 22.1, "is_after_hours": True},
    {"name": "Board Room",        "occupancy_count": 0,
     "temperature_c": 21.0, "is_after_hours": True,
     "manual_override": "heating for 8am meeting"},
]
for zone in zones:
    print(f"  {zone['name']:<22} → {control_hvac(zone)}")
```

Inventory Reorder Decision Engine

Example 2 — Variation

An inventory system that flags items for reorder using compound conditions across multiple fields.

```
def reorder_decision(item: dict) -> tuple[bool, str]:",
    """
    Decide whether to raise a purchase order for an inventory item.
    Returns (should_reorder, reason).
    """
    stock      = item.get("quantity_on_hand", 0)
    threshold  = item.get("reorder_point", 0)
    pending    = item.get("pending_order", False)
    active     = item.get("is_active", True)
    lead_days  = item.get("lead_time_days", 0)
    daily_use  = item.get("daily_usage", 0)

    if not active:
        return False, "Item discontinued"

    if pending:
        return False, "Order already pending"

    # Reorder if stock is at or below threshold
    # OR if projected runout occurs within lead time
    projected_runout = daily_use > 0 and (stock / daily_use) <= lead_days
    if stock <= threshold or projected_runout:
        return True, f"Stock {stock} at/below threshold {threshold}"

    return False, "Stock sufficient"

items = [
    {"sku": "P-001", "quantity_on_hand": 8,  "reorder_point": 20,
     "daily_usage": 3.0, "lead_time_days": 5, "pending_order": False,
"is_active": True},
    {"sku": "P-002", "quantity_on_hand": 50, "reorder_point": 15,
     "daily_usage": 1.0, "lead_time_days": 7, "pending_order": True,
"is_active": True},
]
for item in items:
    reorder, reason = reorder_decision(item)
    flag = "★ REORDER" if reorder else "  ok"
    print(f"{flag}  {item['sku']}: {reason}")
```

Chapter 6 Summary

- Six comparison operators return booleans: == (equal), != (not equal), < (less than), > (greater than), <= (less than or equal), >= (greater than or equal). Chained comparisons like a < b < c are valid and mean (a < b) and (b < c).
- Three logical operators combine booleans: and (both must be True), or (at least one must be True), not (inverts). Python evaluates any value in a boolean context: zero, None, and empty containers are falsy; everything else is truthy.

- Short-circuit evaluation: and stops at the first False operand; or stops at the first True operand. The right-hand side is never evaluated when the result is already determined. This is reliable and can be used deliberately to prevent runtime errors.
- Truth tables enumerate all input combinations and their outputs, providing a definitive specification of what a boolean expression does. Use them to verify that compound conditions match their intended business rules. De Morgan's laws let you transform not(A and B) to (not A) or (not B), and vice versa.
- Compound conditions combine comparisons with logical operators. Assign named boolean variables to individual sub-conditions to make the final compound expression auditable. Break complex conditions across multiple lines inside parentheses for readability.
- Nested conditions check secondary conditions only after an outer condition has passed. Excessive nesting reduces readability; use early-return guard clauses to keep the successful path at the minimal indentation level.
- all(iterable) returns True if every element is truthy; any(iterable) returns True if at least one is. These are the loop-free equivalents of chaining and and or across a collection and are the idiomatic way to check multiple conditions stored in a list or dict.

Guided Exercises

Complete the following exercises in VS Code, IDLE, or your preferred Python environment. Each exercise targets a specific concept from this chapter.

1. Write a server metrics classifier. Given a dictionary containing cpu_pct, mem_pct, error_pct, and latency_ms, write a function that returns "CRITICAL", "DEGRADED", or "HEALTHY" using the following rules: critical if any single metric exceeds its critical threshold; degraded if two or more metrics exceed their warning thresholds; healthy otherwise. Define all thresholds as named constants at the top of the file. Test with at least five different metric combinations, including edge cases on the thresholds.

2. Implement a truth table generator. Write a function truth_table(expr_str, variables) that accepts a string expression (e.g. "A and (B or C)") and a list of variable names, evaluates the expression for every combination of True/False values, and prints a formatted table with a column for each variable and a result column. Demonstrate it with at least two different expressions, one using all three operators.

3. Write a credit pre-qualification function that checks five separate criteria (income, credit score, employment duration, debt-to-income ratio, and absence of defaults). Rather than short-circuiting, it should evaluate all criteria independently and return a dictionary containing: overall approval status (True/False), the result of each individual check, and a list of all failed criteria. Test with at least four applicants, ensuring at least one passes all criteria and at least one fails multiple.

4. Demonstrate short-circuit evaluation with three examples: (1) safe division using and to avoid ZeroDivisionError; (2) safe dictionary key access using and to avoid KeyError; (3) a default value using or to substitute when a value is None, zero, or an empty string. For each, show what would happen without short-circuiting and how the short-circuit expression prevents the error.

5. Implement an inventory reorder decision engine. For each item in a list of dictionaries (with fields: sku, quantity_on_hand, reorder_point, daily_usage, lead_time_days, pending_order, is_active), determine whether a purchase order should be raised. The decision requires: the item is active, no order is already pending, AND either the stock is at or below the reorder point OR the projected stock-out date is within the lead time. Print a formatted report showing each SKU's status and the reason for the decision.

Challenge Tasks

The following tasks are open-ended. No hints are provided. They require you to design a solution, not just apply a pattern.

1. Design and implement a role-based access control (RBAC) system. Define at least five roles (viewer, analyst, editor, manager, admin) and at least eight resources (dashboard, reports, raw_data, user_management, billing, audit_log, api_keys, system_config). Implement a permission matrix as a dictionary mapping (role, resource) pairs to boolean access rights. Write a check_access(user, resource, action) function where action can be "read", "write", or "delete" — with different permission levels per action. Include account status checks (suspended, unverified). Test every role against every resource and print a complete access matrix.

2. Build a smart building zone controller with at least four zone types (open plan office, meeting room, server room, lobby) and at least six configurable conditions per zone (occupancy count, temperature, time of day, CO2 ppm, humidity %, manual override). Each zone type has different comfort targets and control rules. Write control logic that determines HVAC action, lighting level, and ventilation rate for each zone based on its current sensor readings and zone type. Demonstrate the controller with a set of realistic sensor readings that exercises every possible outcome across the zone types.

3. Write a boolean expression auditor. Given a list of business rules expressed as Python lambda functions and a set of test cases with known expected outcomes, the auditor should: evaluate each rule against each test case, report any case where the actual outcome differs from the expected outcome, generate a truth-table-style report for up to three-variable rules, and use De Morgan's law to generate and verify an equivalent negated form of each rule. Include at least six rules covering different combinations of and, or, and not, and at least ten test cases per rule.

CHAPTER 7 Conditional Statements: if, elif, else

Think, Code, Create – A Python Programming Journey

Conditional statements are the mechanism by which a program branches: different code executes depending on which conditions are true at runtime. Without them, every program would produce the same output for every input, which is useful only in the narrowest of circumstances. With them, a single program can classify transactions, enforce access rules, apply tiered pricing, route requests, and respond intelligently to the full range of inputs it might receive.

This chapter covers the full conditional toolkit in Python: the basic if statement, the if-else two-branch structure, elif chains for multiple mutually exclusive cases, compound conditions, nested conditions, sequences of independent checks, and robust input validation. Every section includes real-world examples that show these tools in the kinds of contexts you will actually encounter in practice.

7.1 The if Statement

The if statement executes a block of code only when a specified condition evaluates to True. If the condition is False, the block is skipped entirely and execution continues at the next statement after the block. This is the simplest form of branching: a single test with a single consequence.

The block that belongs to the if statement is defined by indentation — every line of the block must be indented one level further than the if keyword itself. Python uses this indentation as the block delimiter instead of braces. A line at the same indentation level as the if keyword is outside the block and executes unconditionally.

Basic Usage

Example 1 — Introduction

if statements in a payment processing context: each check fires independently of the others.

```
transaction_amount = 4_250.00
account_balance    = 3_800.00
is_verified        = True
LARGE_TRANSACTION_THRESHOLD = 3_000

if not is_verified:
    print("BLOCKED: account not verified")

if transaction_amount > account_balance:
    print("ALERT: transaction exceeds available balance")

if transaction_amount > LARGE_TRANSACTION_THRESHOLD:
    print(f"NOTICE: large transaction flagged for review
($ {transaction_amount:,.2f})")
```

```
# This line always executes -- it is outside all if blocks
print("Transaction processing complete")
```

Notice that each if statement is independent: all three are evaluated, and any whose condition is True executes its block. This is sequential evaluation, not a chain — a point that becomes important in section 7.6 when contrasted with if-elif-else.

Indentation and Block Membership

Example 2 — Variation

Demonstrating block membership: indented lines belong to the if; unindented lines do not.

```
response_time_ms = 1_450
WARN_THRESHOLD   = 500

if response_time_ms > WARN_THRESHOLD:
    # All three lines below are inside the if block
    print(f"Slow response detected: {response_time_ms}ms")
    print(f"Threshold: {WARN_THRESHOLD}ms")
    print("Check database connection and query plans")

# This line is outside the if block -- always executes
print("Health check complete")
```

if with Expressions

The condition in an if statement can be any expression that evaluates to a boolean: a comparison, a logical expression, a function call, or even a simple variable (using Python's truthiness rules). All of the following are valid if conditions:

Example 3 — Real-world application

Various condition forms: comparison, membership test, function result, truthiness, compound.

```
status_code  = 404
error_codes  = {400, 401, 403, 404, 500, 502, 503}
pending_jobs = ["job-001", "job-002"]

if status_code >= 400:                      # comparison
    print("HTTP error:", status_code)

if status_code in error_codes:              # membership test
    print("Known error code")

if pending_jobs:                            # truthiness (non-empty list)
    print(f"{len(pending_jobs)} job(s) pending")

if status_code >= 500 and status_code < 600:  # compound
    print("Server-side error -- check application logs")
```

7.2 The if-else Statement

The if-else statement handles exactly two mutually exclusive branches: one that executes when the condition is True, and one that executes when the condition is False. Exactly one branch always executes — the two possibilities are exhaustive and disjoint. This makes if-else appropriate whenever there are precisely two outcomes and every path through the code must produce a result.

Two-Branch Decisions

Example 1 — Introduction

if-else for binary routing decisions in a transaction processing system.

```
def route_transaction(transaction: dict) -> str:
    amount    = transaction["amount"]
    tx_type   = transaction["type"]
    risk_flag = transaction.get("risk_flag", False)

    if risk_flag:
        return "HOLD: flagged for manual review"
    else:
        return f"PROCESS: {tx_type} of ${amount:,.2f}"

transactions = [
    {"amount": 850.00,  "type": "purchase",  "risk_flag": False},
    {"amount": 12_000,  "type": "transfer",  "risk_flag": True},
    {"amount": 34.99,   "type": "purchase",  "risk_flag": False},
]
for tx in transactions:
    print(route_transaction(tx))
```

if-else as an Expression: Conditional Expressions

Python supports an inline conditional expression (sometimes called a ternary operator): value_if_true if condition else value_if_false. This is not a statement but an expression, so it can be used anywhere a value is expected — in an assignment, as a function argument, or inside an f-string. Use it for simple, readable two-outcome assignments; prefer a full if-else block when the logic is complex or when either branch involves multiple steps.

Example 2 — Variation

Conditional expressions for concise single-value decisions in a reporting context.

```
stock     = 0
threshold = 20

# Conditional expression: cleaner than a two-line if-else for a single value
status = "OUT OF STOCK" if stock == 0 else ("LOW" if stock < threshold else
"OK")
print(f"Stock status: {status}")

# Inside an f-string
error_count = 7
severity = "warning" if error_count < 10 else "critical"
print(f"Errors: {error_count} ({severity})")
```

```
# As a function argument
records = [{"active": True}, {"active": False}, {"active": True}]
active_count = sum(1 if r["active"] else 0 for r in records)
print(f"Active records: {active_count} of {len(records)}")
```

Avoiding Common if-else Mistakes

Two mistakes appear repeatedly with if-else. The first is the accidental fall-through: placing code that should be inside the else block at the same indentation as the if, so it always executes. The second is using = (assignment) instead of == (comparison) in the condition — Python raises a SyntaxError for this, but the error message can be confusing.

Example 3 — Real-world application

Demonstrating correct vs incorrect block membership in an alert routing function.

```
# WRONG: second print is outside the else block -- always executes
def alert_wrong(is_critical):
    if is_critical:
        print("Page on-call engineer")
    else:
        print("Log warning to dashboard")
    print("Send email notification")  # runs for BOTH branches

# CORRECT: notification only sent when not critical
def alert_correct(is_critical):
    if is_critical:
        print("Page on-call engineer")
    else:
        print("Log warning to dashboard")
        print("Send email notification")  # indented into else block

print("--- Wrong ---")
alert_wrong(True)
print("--- Correct ---")
alert_correct(True)
```

7.3 The elif Statement

elif (else-if) chains multiple mutually exclusive branches together. Python evaluates each condition in order from top to bottom and executes the first block whose condition is True, then skips all remaining branches. If no condition is True and an else block is present, the else block executes. If no else is provided and no condition matches, nothing executes.

elif is appropriate when you have three or more mutually exclusive cases. It is more efficient than a sequence of independent if statements (section 7.6) because it stops evaluating once a match is found. Order matters: conditions are checked in sequence, so place the most specific or most likely condition first.

Tax Bracket Classification

Example 1 — Introduction

elif chain applying Australian income tax brackets to compute tax payable.

```
# Australian 2023-24 resident tax brackets (simplified)
def calculate_tax(taxable_income: float) -> float:
    if taxable_income <= 18_200:
        return 0.0
    elif taxable_income <= 45_000:
        return (taxable_income - 18_200) * 0.19
    elif taxable_income <= 120_000:
        return 5_092 + (taxable_income - 45_000) * 0.325
    elif taxable_income <= 180_000:
        return 29_467 + (taxable_income - 120_000) * 0.37
    else:
        return 51_667 + (taxable_income - 180_000) * 0.45

incomes = [15_000, 42_000, 85_000, 135_000, 220_000]
print(f"{"Income":>12}  {"Tax":>12}  {"Effective rate":>14}")
print("-" * 42)
for income in incomes:
    tax  = calculate_tax(income)
    rate = tax / income * 100 if income > 0 else 0
    print(f"${income:>11,}  ${tax:>11,.2f}  {rate:>12.1f}%")
```

Priority Routing

Example 2 — Variation

elif used to route support tickets to the correct team based on priority classification.

```
def route_ticket(ticket: dict) -> str:
    priority  = ticket.get("priority", "low")
    category  = ticket.get("category", "general")

    if priority == "critical":
        return "IMMEDIATE: page on-call engineer"
    elif priority == "high" and category == "security":
        return "URGENT: security team + engineering lead"
    elif priority == "high":
        return "HIGH: senior engineering team (SLA: 4h)"
    elif priority == "medium":
        return "NORMAL: standard queue (SLA: 24h)"
    else:
        return "LOW: backlog queue (SLA: 72h)"

tickets = [
    {"priority": "critical", "category": "outage"},
    {"priority": "high",     "category": "security"},
    {"priority": "high",     "category": "performance"},
    {"priority": "medium",   "category": "billing"},
    {"priority": "low",      "category": "feature"},
]
for t in tickets:
    print(f"[{t['priority'].upper():<8}] {route_ticket(t)}")
```

Why Order Matters in elif Chains

Example 3 — Real-world application

Showing how incorrect elif ordering silently produces wrong results in a tier classifier.

```
usage_gb = 85

# WRONG ORDER: the first match wins, so "Standard" catches everything >= 50
# and the "Premium" branch is never reached
if usage_gb >= 50:
    tier = "Standard"
elif usage_gb >= 100:   # never reached for usage >= 50
    tier = "Premium"
else:
    tier = "Free"
print(f"Wrong:   {tier}")   # prints Standard for usage_gb=85

# CORRECT ORDER: most restrictive first
if usage_gb >= 100:
    tier = "Premium"
elif usage_gb >= 50:
    tier = "Standard"
else:
    tier = "Free"
print(f"Correct: {tier}")   # prints Standard correctly
```

7.4 Using elif with Complex Conditions

Each branch in an if-elif-else chain can use any valid boolean expression as its condition — including compound expressions built with and, or, and not, chained comparisons, function calls, and membership tests. Complex conditions allow a single elif chain to handle a multi-dimensional decision space without nesting.

When combining multiple operators in a condition, use parentheses to make the grouping explicit. A reader should not need to apply the precedence table to understand which sub-expressions are combined with which operators.

Compound Conditions in elif Branches

Example 1 — Introduction

Network packet classification using compound conditions to route traffic by protocol and port.

```
WELL_KNOWN_HTTP   = {80, 443}
WELL_KNOWN_DB     = {5432, 3306, 1521}   # PostgreSQL, MySQL, Oracle
MONITORING_PORTS = {9090, 9100, 8080}    # Prometheus, node_exporter, etc.

def classify_packet(protocol: str, port: int, is_internal: bool) -> str:
    if protocol == "TCP" and port in WELL_KNOWN_HTTP:
        return "WEB: route to HTTP load balancer"
    elif protocol == "TCP" and port in WELL_KNOWN_DB and is_internal:
        return "DB: route to database cluster (internal only)"
    elif protocol == "TCP" and port in WELL_KNOWN_DB and not is_internal:
        return "REJECT: database ports not exposed externally"
    elif port in MONITORING_PORTS and is_internal:
        return "MONITOR: route to metrics collector"
```

```
    elif protocol == "UDP":
        return "UDP: route to UDP handler"
    else:
        return "DEFAULT: route to general pool"

tests = [
    ("TCP", 443,  True),
    ("TCP", 5432, True),
    ("TCP", 5432, False),
    ("TCP", 9090, True),
    ("UDP", 53,   False),
]
for proto, port, internal in tests:
    direction = "internal" if internal else "external"
    print(f"  {proto} {port:>5} {direction:<9}  {classify_packet(proto, port,
internal)}")
```

Combining and and or in Conditions

When a branch condition uses both and and or, apply parentheses even when the default precedence produces the correct result — because a future reader may not immediately remember that and binds tighter than or. Parentheses cost nothing and prevent misreadings.

Example 2 — Variation

Risk classification using and-or compound conditions with explicit parentheses.

```
def classify_risk(transaction: dict) -> str:
    amount        = transaction["amount"]
    country       = transaction["country"]
    is_new_payee  = transaction.get("is_new_payee", False)
    is_overnight  = transaction.get("is_overnight", False)

    HIGH_RISK_COUNTRIES = {"IR", "KP", "SY"}

    if country in HIGH_RISK_COUNTRIES:
        return "HIGH: restricted jurisdiction"
    elif amount > 50_000 and (is_new_payee or is_overnight):
        return "HIGH: large amount to new or overnight payee"
    elif amount > 10_000 and is_new_payee:
        return "MEDIUM: first transaction to this payee"
    elif amount > 50_000:
        return "MEDIUM: large transaction, known payee"
    else:
        return "LOW: routine transaction"

txns = [
    {"amount": 75_000, "country": "AU", "is_new_payee": True,  "is_overnight":
False},
    {"amount": 12_000, "country": "AU", "is_new_payee": True,  "is_overnight":
False},
    {"amount": 500,    "country": "KP", "is_new_payee": False, "is_overnight":
False},
]
for t in txns:
    print(f"  ${t['amount']:>8,}  {classify_risk(t)}")
```

7.5 Nested if Statements

A nested if statement places a conditional block inside another conditional block. The inner condition is only evaluated when the outer condition is True, which makes it natural for hierarchical decision logic: first establish a broad context, then make a more specific decision within that context.

The principal risk of nesting is that readability degrades quickly with depth. More than two or three levels of nesting is a signal to refactor — either by flattening with guard clauses (early return) or by extracting the inner logic into a named function.

Two-Level Nesting

Example 1 — Introduction

Nested conditions in an API request dispatcher: outer checks authentication, inner routes by method.

```
def dispatch(request: dict) -> str:
    token = request.get("token")
    method = request.get("method")
    path   = request.get("path")

    if token and validate_token(token):
        # Authenticated: route by HTTP method
        if method == "GET":
            return f"Read resource: {path}"
        elif method in ("POST", "PUT", "PATCH"):
            return f"Write resource: {path}"
        elif method == "DELETE":
            return f"Delete resource: {path}"
        else:
            return f"405 Method Not Allowed: {method}"
    else:
        return "401 Unauthorized: valid token required"

def validate_token(token: str) -> bool:
    return token.startswith("Bearer ") and len(token) > 10

requests = [
    {"token": "Bearer abc123xyz", "method": "GET",    "path": "/api/orders"},
    {"token": "Bearer abc123xyz", "method": "DELETE", "path":
"/api/orders/99"},
    {"token": None,               "method": "GET",    "path": "/api/orders"},
]
for r in requests:
    print(f"  {dispatch(r)}")
```

Refactoring Nested Conditions with Guard Clauses

Guard clauses check failure conditions first and return immediately, leaving the main success path with minimal nesting. This pattern — also called early return — typically produces code that is easier to read and test than an equivalent nested structure.

Example 2 — Variation

The same order fulfilment logic written with deep nesting versus flat guard clauses.

```
# Deep nesting -- hard to follow
def fulfil_nested(order):
    if order:
        if order.get("paid"):
            if order.get("stock_reserved"):
                if order.get("address"):
                    return "Dispatching order"
                else:
                    return "Missing delivery address"
            else:
                return "Stock not reserved"
        else:
            return "Payment not received"
    else:
        return "No order provided"

# Flat guard clauses -- same logic, much easier to read
def fulfil_flat(order):
    if not order:            return "No order provided"
    if not order.get("paid"):         return "Payment not received"
    if not order.get("stock_reserved"): return "Stock not reserved"
    if not order.get("address"):      return "Missing delivery address"
    return "Dispatching order"

test = {"paid": True, "stock_reserved": True, "address": None}
print(fulfil_nested(test))
print(fulfil_flat(test))
```

When Nesting Is Appropriate

Example 3 — Real-world application

IoT device command dispatcher: category determines outer branch, command determines inner action.

```
def handle_command(device_type: str, command: str, params: dict) -> str:
    if device_type == "hvac":
        if command == "set_temp":
            target = params.get("celsius", 21)
            return f"HVAC: setting target temperature to {target}°C"
        elif command == "set_mode":
            mode = params.get("mode", "auto")
            return f"HVAC: switching to {mode} mode"
        else:
            return f"HVAC: unknown command '{command}'"
    elif device_type == "lighting":
        if command == "set_level":
            level = max(0, min(100, params.get("pct", 100)))
            return f"Lighting: dimming to {level}%"
        elif command == "set_colour":
            return f"Lighting: colour set to {params.get('kelvin', 4000)}K"
        else:
            return f"Lighting: unknown command '{command}'"
    else:
        return f"Unknown device type: '{device_type}'"
```

```
print(handle_command("hvac",     "set_temp",   {"celsius": 22}))
print(handle_command("lighting", "set_level",  {"pct": 60}))
print(handle_command("sensor",   "read",       {}))
```

7.6 Multiple if Statements in Sequence

When multiple if statements appear one after another — without elif — each is evaluated independently. Every condition whose test is True fires its block, regardless of whether a previous condition also fired. This is fundamentally different from an if-elif chain, where only the first matching branch executes.

Use sequential if statements when the conditions are independent and multiple may legitimately apply simultaneously. Use if-elif when the conditions are mutually exclusive and at most one should fire. Confusing these two patterns is one of the most common sources of subtle logic bugs.

Independent Checks

Example 1 — Introduction

A multi-metric health dashboard: each metric checked independently, multiple alerts can fire.

```
def check_service_health(metrics: dict) -> list[str]:
    """
    All checks are independent -- multiple alerts can fire simultaneously.
    Returns a list of alert messages.
    """
    alerts = []

    cpu_pct     = metrics.get("cpu_pct",    0)
    mem_pct     = metrics.get("mem_pct",    0)
    disk_pct    = metrics.get("disk_pct",   0)
    error_rate  = metrics.get("error_rate", 0)
    latency_ms  = metrics.get("latency_ms", 0)

    # Each if is independent -- all are evaluated
    if cpu_pct >= 90:
        alerts.append(f"CPU critical: {cpu_pct}%")
    if mem_pct >= 85:
        alerts.append(f"Memory high: {mem_pct}%")
    if disk_pct >= 95:
        alerts.append(f"Disk critical: {disk_pct}%")
    if error_rate >= 5:
        alerts.append(f"High error rate: {error_rate}%")
    if latency_ms >= 2_000:
        alerts.append(f"Latency degraded: {latency_ms}ms")

    return alerts if alerts else ["All metrics nominal"]

metrics = {"cpu_pct": 93, "mem_pct": 88, "disk_pct": 60,
           "error_rate": 1.2, "latency_ms": 3_400}
for alert in check_service_health(metrics):
    print(f"  • {alert}")
```

Sequential if vs elif: The Distinction in Practice

Example 2 — Variation

The same input processed by sequential if (wrong for mutually exclusive cases) vs elif (correct).

```
order_status = "shipped"

# WRONG: sequential if -- all matching conditions fire
# "shipped" does not match "pending", but the elif intent was for only one to
fire
print("Sequential if:")
if order_status == "pending":
    print("  Awaiting payment confirmation")
if order_status == "processing":
    print("  Being prepared for dispatch")
if order_status == "shipped":
    print("  En route to customer")
if order_status == "shipped" or order_status == "processing":
    print("  Send shipping notification")  # fires for TWO statuses!

# CORRECT: elif -- only the first match executes
print("elif chain:")
if order_status == "pending":
    print("  Awaiting payment confirmation")
elif order_status == "processing":
    print("  Being prepared for dispatch")
    print("  Send shipping notification")
elif order_status == "shipped":
    print("  En route to customer")
```

Accumulating Results Across Independent Checks

Example 3 — Real-world application

A data quality checker applying multiple independent validation rules and collecting all failures.

```
def validate_record(record: dict) -> list[str]:
    """Check all rules independently and return all failures."""
    issues = []

    if not record.get("customer_id"):
        issues.append("Missing customer_id")

    amount = record.get("amount")
    if amount is None:
        issues.append("Missing amount")
    elif not isinstance(amount, (int, float)) or amount <= 0:
        issues.append(f"Invalid amount: {amount!r}")

    if not record.get("currency") in ("AUD", "USD", "EUR", "GBP"):
        issues.append(f"Unsupported currency: {record.get('currency')!r}")

    if not record.get("timestamp"):
        issues.append("Missing timestamp")

    return issues
```

```
records = [
    {"customer_id": "C001", "amount": 49.95, "currency": "AUD", "timestamp":
"2024-07-15"},
    {"customer_id": "",     "amount": -10,   "currency": "BTC", "timestamp":
None},
]
for r in records:
    problems = validate_record(r)
    status   = "OK" if not problems else f"FAIL ({len(problems)} issue(s))"
    print(f"  {status}: {problems if problems else r['customer_id']}")
```

7.7 Input Validation with Conditional Statements

Input validation is the discipline of verifying that a value meets all requirements before allowing it to proceed through the program's logic. It is not optional: any value that originates from outside the program — terminal input, file contents, API responses, database reads — must be treated as potentially incorrect. Validation with conditional statements is the mechanism that enforces correctness at the boundary between the outside world and the program's internals.

The Core Validation Pattern

The standard pattern for interactive input validation is: prompt → try conversion → test constraints → return on success, loop on failure. Encapsulate this in a dedicated function so the calling code remains clean. Use try/except for type conversion (never str.isdigit() — it fails on floats, scientific notation, and negative numbers). Test value constraints with if statements after a successful conversion.

Example 1 — Introduction

Reusable validated-input functions for integers and floats with constraint checking.

```
def prompt_int(prompt: str, min_val: int = None, max_val: int = None) -> int:
    while True:
        raw = input(prompt)
        try:
            value = int(raw)
        except ValueError:
            print(f"  Expected a whole number, got: {raw!r}")
            continue

        if min_val is not None and value < min_val:
            print(f"  Must be at least {min_val}.")
        elif max_val is not None and value > max_val:
            print(f"  Must be at most {max_val}.")
        else:
            return value

# Usage
port     = prompt_int("Port number (1-65535): ", 1, 65_535)
timeout  = prompt_int("Timeout in seconds (5-300): ", 5, 300)
print(f"Connecting on port {port} with {timeout}s timeout")
```

Why Not isdigit()?

The str.isdigit() method is frequently suggested for numeric validation but is almost always the wrong choice. It returns False for any string containing a decimal point, a sign character, or scientific notation — so "3.14", "-5", and "1e6" all fail. It returns True for strings containing non-ASCII digit characters such as ² (superscript two). The try/except pattern covers all valid numeric forms and is the correct approach.

```
# isdigit() failures:
print("3.14".isdigit())    # False -- but float("3.14") works fine
print("-5".isdigit())      # False -- but int("-5") works fine
print("1e6".isdigit())     # False -- but float("1e6") works fine
print("\u00b2".isdigit()) # True  -- but int("\u00b2") fails!

# Correct approach for all valid integers:
def is_valid_int(s: str) -> bool:
    try:
        int(s)
        return True
    except ValueError:
        return False

print(is_valid_int("-5"))  # True
print(is_valid_int("3.14")) # False (correct for integer context)
```

Validating String Fields and Choice Lists

Example 2 — Variation

String field validation: non-empty, length bounds, and membership in an allowed set.

```
VALID_ENVIRONMENTS = {"dev", "staging", "prod"}
VALID_LOG_LEVELS   = {"DEBUG", "INFO", "WARNING", "ERROR", "CRITICAL"}

def prompt_nonempty(prompt: str, min_len: int = 1, max_len: int = None) ->
str:
    while True:
        val = input(prompt).strip()
        if len(val) < min_len:
            print(f"  Must be at least {min_len} character(s).")
        elif max_len and len(val) > max_len:
            print(f"  Must be at most {max_len} character(s).")
        else:
            return val

def prompt_choice(prompt: str, choices: set) -> str:
    sorted_choices = sorted(choices)
    while True:
        val = input(f"{prompt} ({'/'.join(sorted_choices)}): ").strip()
        if val in choices:
            return val
        print(f"  Must be one of: {sorted_choices}")

env       = prompt_choice("Environment", VALID_ENVIRONMENTS)
log_level = prompt_choice("Log level",   VALID_LOG_LEVELS)
service   = prompt_nonempty("Service name: ", min_len=3, max_len=50)
```

```
print(f"Deploying {service} to {env} with log level {log_level}")
```

Validating Structured Input

Example 3 — Real-world application

Date string validation with format checking and logical range constraints.

```
from datetime import datetime, date

def prompt_date(prompt: str, fmt: str = "%d/%m/%Y",
                min_date: date = None,
                max_date: date = None) -> date:
    while True:
        raw = input(f"{prompt} ({fmt.replace('%d', 'DD').replace('%m',
'MM').replace('%Y', 'YYYY')}): ").strip()
        try:
            parsed = datetime.strptime(raw, fmt).date()
        except ValueError:
            print(f"  Invalid format. Use {fmt}.")
            continue

        if min_date and parsed < min_date:
            print(f"  Date must be on or after {min_date.strftime(fmt)}.")
        elif max_date and parsed > max_date:
            print(f"  Date must be on or before {max_date.strftime(fmt)}.")
        else:
            return parsed

today    = date.today()
start_dt = prompt_date("Start date", min_date=today)
end_dt   = prompt_date("End date",   min_date=start_dt)
print(f"Reporting period: {start_dt} to {end_dt} ({(end_dt - start_dt).days}
days)")
```

Chapter 7 Summary

- The if statement executes its block only when the condition is True; the block is defined by indentation. Multiple statements in the block are all indented one level further than the if keyword. Lines at the same indentation level as if are outside the block and always execute.
- if-else provides exactly two mutually exclusive branches: the if block runs on True, the else block runs on False. Exactly one branch always executes. The conditional expression (value_if_true if condition else value_if_false) is the one-line equivalent for simple value selection.
- elif chains multiple mutually exclusive conditions. Python evaluates from top to bottom and executes the first matching branch, then skips the rest. Order matters: place more specific or restrictive conditions before broader ones. If no condition matches and no else is provided, nothing executes.
- Each branch in an if-elif-else chain can use any boolean expression, including compound expressions built with and, or, not, chained comparisons, membership tests, and function calls. Use parentheses to make grouping explicit whenever more than two operators are combined.

- Nested if statements check an inner condition only when the outer condition is True. Limit nesting to two or three levels for readability. Use guard clauses (early return) to flatten nested structures: check each failure condition first and return immediately, leaving the successful path with minimal indentation.
- Sequential independent if statements each evaluate unconditionally — multiple blocks can fire for the same input. Use this pattern when conditions are independent. Use if-elif when conditions are mutually exclusive. Confusing the two is a frequent source of logic bugs.
- For input validation, always use try/except for type conversion — not str.isdigit(), which fails on floats, negatives, and scientific notation. After a successful conversion, use if statements to check value constraints. Encapsulate validation in reusable functions that loop until valid input is received.

Guided Exercises

Complete the following exercises in VS Code, IDLE, or your preferred Python environment. Each exercise targets a specific concept from this chapter.

1. Write a function calculate_tax(taxable_income) that implements a five-bracket income tax schedule. Define all bracket boundaries and rates as named constants at the top of the file. Test with at least eight income values covering all brackets and boundary conditions (exactly on a bracket threshold, just above, and just below). Print a formatted table showing income, tax payable, and effective rate for each test value.

2. Build a support ticket router. A ticket has a priority ("critical", "high", "medium", "low") and a category (e.g. "security", "performance", "billing", "general"). Write an elif chain that routes each combination to the correct team and SLA. Include at least six distinct routing rules using compound conditions. Test with a set of tickets covering every combination, and print each ticket's routing decision.

3. Implement a data record validator using sequential independent if statements. The validator should check at least six fields independently (e.g. customer_id non-empty, amount positive float, currency in approved list, timestamp present, country code valid, quantity a positive integer). Every failing check should add a message to an errors list. Return the list of all errors, not just the first one. Test with at least three records: one fully valid, one with a single error, one with multiple errors.

4. Write a prompt_date() function that collects a date string from the user, validates the format with try/except and datetime.strptime(), checks that the date falls within an acceptable range (e.g. not in the past, not more than two years in the future), and loops until valid input is received. Extend it to collect two dates — a start and an end — where the end must be after the start. Print the number of days in the resulting range.

5. Demonstrate the if vs elif distinction with a concrete example. Write two versions of the same function: one using sequential if statements, one using an elif chain. Choose a domain where the difference matters (e.g. order status processing, severity classification). Show at least three inputs where the two versions produce different outputs, and explain in comments why each difference occurs.

Challenge Tasks

The following tasks are open-ended. No hints are provided. They require you to design a solution, not just apply a pattern.

1. Design and implement a complete freight rate calculator. Given a shipment's weight (kg), volume (m^3), destination zone (domestic/international), service level (standard/express/overnight), and declared value (for insurance), calculate: the chargeable weight (greater of actual weight and volumetric weight at 250kg/m^3), the base freight rate from a rate table, any applicable surcharges (fuel surcharge, remote area, oversize, weekend delivery), insurance (0.5% of declared value, minimum $15), and GST (10% on freight + surcharges, not insurance). Use elif chains for rate tiers, sequential if statements for independent surcharges, and guard clauses for input validation. Print a fully itemised quote.

2. Build a command-line deployment tool for a hypothetical application. It should: collect and validate environment (dev/staging/prod), service name, version tag (semantic version format x.y.z), and confirmation for production deployments; apply different deployment rules per environment (e.g. prod requires approval flag and disallows pre-release versions); use nested conditions to enforce environment-specific constraints; and print a detailed deployment plan showing all parameters and any warnings. Include at least five validation rules and at least three environment-specific behavioural differences.

3. Implement a multi-step diagnostic tool. The tool should ask a series of yes/no questions (at least eight) and use a conditional decision tree to narrow down the likely cause of a problem in a domain of your choice (e.g. network connectivity, application startup failure, hardware fault). Each path through the tree should end at a specific diagnosis with recommended actions. The tree must have at least four distinct terminal diagnoses, at least one path requiring three or more correct responses, and at least one branch that loops back to re-ask a question. Document the decision tree logic separately in comments before writing the code.

CHAPTER 8 **Looping with while**

Think, Code, Create – A Python Programming Journey

> The while loop is Python's mechanism for conditional repetition: execute a block of code for as long as a condition remains True, checking the condition before each iteration. Unlike the for loop, which iterates over a known sequence, the while loop continues until something in the world changes — a value updates, a flag is set, a resource becomes available, or a user provides acceptable input. This makes it the natural choice for interactive programs, polling systems, retry logic, and any process where the number of iterations is not known in advance.
>
> This chapter covers the full while loop toolkit: basic structure and exit conditions, multiple conditions, infinite loops with break, continue and pass for flow control within the loop, flag variables and sentinel values, data-driven processing, and the while-else clause. All examples are drawn from practical domains — validation, monitoring, retry logic, and data processing — rather than abstract counting exercises.

8.1 Introduction to Looping with while

A while loop evaluates its condition before each iteration. If the condition is True, the loop body executes, then the condition is evaluated again. This cycle continues until the condition is False, at which point execution jumps to the first statement after the loop. If the condition is False before the loop is even entered, the loop body never executes.

The fundamental contract of a while loop: the loop body must eventually cause the condition to become False, or the loop must contain a break statement. A loop where neither happens is an infinite loop — intentional in server and monitoring contexts, but a defect if accidental.

Condition-Driven Loops

Example 1 — Introduction

A retry loop that attempts a network connection up to a maximum number of times.

```
import time

MAX_RETRIES = 5
RETRY_DELAY = 2.0   # seconds

def attempt_connection(host: str) -> bool:
    """Simulates a connection attempt. Returns True on success."""
    import random
    return random.random() > 0.6  # 40% failure rate for demonstration

host    = "db-prod-01.internal"
retries = 0
connected = False
```

```
while retries < MAX_RETRIES and not connected:
    retries += 1
    print(f"Attempt {retries}/{MAX_RETRIES}: connecting to {host}...")
    connected = attempt_connection(host)
    if not connected and retries < MAX_RETRIES:
        print(f"  Failed. Retrying in {RETRY_DELAY}s...")
        time.sleep(RETRY_DELAY)

if connected:
    print(f"Connected after {retries} attempt(s).")
else:
    print(f"Failed to connect after {MAX_RETRIES} attempts. Aborting.")
```

The condition retries < MAX_RETRIES and not connected ensures the loop stops on either success or exhausting the retry budget. Both the counter and the flag are updated inside the loop body, guaranteeing the loop will eventually terminate.

State-Driven Loops

Many while loops track a piece of state that evolves as the loop runs. The condition tests that state. When the state reaches a terminal value, the loop exits naturally without needing a break.

Example 2 — Variation

A stock depletion simulator: loop runs until stock reaches zero or a reorder threshold.

```
REORDER_POINT  = 20
DAILY_USAGE    = 7
opening_stock  = 100

stock      = opening_stock
day        = 0

while stock > REORDER_POINT:
    day   += 1
    stock -= DAILY_USAGE
    stock  = max(stock, 0)    # cannot go below zero
    print(f"Day {day:>3}: stock = {stock:>4} units")

print(f"\nReorder triggered after {day} day(s).")
print(f"Remaining stock: {stock} units")
```

Loop Anatomy

Every correctly written while loop has four components: initialisation (setting up variables before the loop), the condition (tested before each iteration), the body (the work performed each iteration), and the update (modifying the state that the condition tests). Missing the update is the most common cause of unintentional infinite loops.

Example 3 — Real-world application

A payment plan calculator: all four loop components labelled in context.

```
# INITIALISATION
loan_balance   = 25_000.00
```

```
monthly_rate   = 0.0589 / 12
monthly_payment = 480.00
month          = 0
total_interest = 0.0

# CONDITION: loop while balance remains
while loan_balance > 0:
    month    += 1                               # UPDATE (month counter)
    interest  = loan_balance * monthly_rate     # BODY: compute interest
    principal = min(monthly_payment - interest, loan_balance)
    loan_balance -= principal                   # UPDATE (balance)
    total_interest += interest
    if month <= 3 or loan_balance < monthly_payment:
        print(f"Month {month:>3}: interest=${interest:.2f} "
              f"principal=${principal:.2f} balance=${loan_balance:,.2f}")

print(f"\nPaid off in {month} months. Total interest: ${total_interest:,.2f}")
```

8.2 The while Loop with Multiple Conditions

A while loop's condition can be any boolean expression, including compound expressions built with and, or, and not. This allows loops to exit when any one of several conditions is met, or only when all conditions are simultaneously satisfied. The logical structure of the condition directly controls the loop's exit behaviour.

Exiting on Any Condition: or

When the loop condition uses and to combine sub-conditions, the loop continues only while all are True — it exits as soon as any one becomes False. This is the pattern for "run until something goes wrong or a limit is reached":

Example 1 — Introduction

A data ingestion loop that exits when either the record limit is reached or an error occurs.

```
MAX_RECORDS = 1_000

records_ingested = 0
error_occurred   = False

while records_ingested < MAX_RECORDS and not error_occurred:
    try:
        record = fetch_next_record()    # hypothetical data source function
        process(record)
        records_ingested += 1
    except Exception as e:
        print(f"Error at record {records_ingested + 1}: {e}")
        error_occurred = True

print(f"Ingested {records_ingested} records.")
if error_occurred:
    print("Ingestion halted due to error.")
```

Continuing While Any Condition Holds: or

When the condition uses or, the loop continues while at least one sub-condition is True — it exits only when all are False simultaneously. This is useful when multiple independent criteria must all clear before the loop should stop:

Example 2 — Variation

A queue drainer that runs while any worker has pending tasks or the timeout has not expired.

```
import time

workers        = {"worker-1": 5, "worker-2": 3, "worker-3": 8}  # tasks
remaining
timeout_secs   = 30
start_time     = time.time()

def any_pending(workers):
    return any(count > 0 for count in workers.values())

def elapsed(start):
    return time.time() - start

# Loop while tasks remain OR timeout has not expired
while any_pending(workers) and elapsed(start_time) < timeout_secs:
    for worker, count in workers.items():
        if count > 0:
            workers[worker] -= 1
    print(f"[{elapsed(start_time):.1f}s] Remaining: {workers}")
    time.sleep(0.1)

if not any_pending(workers):
    print("All tasks completed.")
else:
    print(f"Timeout after {timeout_secs}s with tasks remaining: {workers}")
```

Complex Multi-Condition Exit Logic

Example 3 — Real-world application

A circuit breaker loop: continues while below error threshold AND within time window AND not manually tripped.

```
import time

MAX_ERRORS    = 5
WINDOW_SECS   = 60

error_count   = 0
tripped       = False
start_time    = time.time()

def is_within_window():
    return time.time() - start_time < WINDOW_SECS

def call_service() -> bool:
    import random
```

```
    return random.random() > 0.3    # 30% error rate

while error_count < MAX_ERRORS and is_within_window() and not tripped:
    success = call_service()
    if not success:
        error_count += 1
        print(f"Error {error_count}/{MAX_ERRORS}")
        if error_count >= MAX_ERRORS:
            tripped = True

if tripped:
    print("CIRCUIT OPEN: too many errors. Requests blocked.")
else:
    print("CIRCUIT CLOSED: operating normally.")
```

8.3 Infinite Loops and break

A while True: loop runs indefinitely because True never becomes False. The only way to exit it is with a break statement, a return from inside a function, or an exception that propagates out. Infinite loops are not defects — they are the standard pattern for event-driven programs, server request handlers, interactive menu systems, and monitoring loops where the program is intended to run until explicitly stopped.

break exits the innermost enclosing loop immediately, transferring control to the first statement after the loop. It should be used when the exit condition is more naturally expressed as a positive test inside the loop body than as a negative condition in the loop header.

Interactive Menu Loop

Example 1 — Introduction

A command-line menu that runs until the user explicitly chooses to exit.

```
def show_menu():
    print("\n=== Inventory Manager ===")
    print("[1] View stock levels")
    print("[2] Record receipt")
    print("[3] Record dispatch")
    print("[4] Generate report")
    print("[Q] Quit")

def get_stock():     return {"Widget A": 142, "Widget B": 23, "Widget C": 87}
def record_receipt(): print("  Receipt recorded.")
def record_dispatch(): print("  Dispatch recorded.")
def generate_report(): print("  Report generated.")

while True:
    show_menu()
    choice = input("> ").strip().lower()
    if choice == "1":
        for item, qty in get_stock().items():
            print(f"  {item:<12} {qty:>5} units")
    elif choice == "2":  record_receipt()
    elif choice == "3":  record_dispatch()
    elif choice == "4":  generate_report()
```

```
    elif choice == "q":
        print("Goodbye.")
        break
    else:
        print("  Unrecognised option.")
```

Monitoring Loop with break on Critical State

Example 2 — Variation

A service monitor that polls indefinitely but breaks out on a critical alert.

```
import time, random

POLL_INTERVAL = 5   # seconds
ALERT_THRESHOLD = 90

def get_cpu_usage():
    return random.randint(40, 99)

print("Starting service monitor. Press Ctrl+C to stop.")

while True:
    cpu = get_cpu_usage()
    timestamp = time.strftime("%H:%M:%S")

    if cpu >= ALERT_THRESHOLD:
        print(f"[{timestamp}] CRITICAL: CPU={cpu}% -- halting monitor")
        break
    elif cpu >= 75:
        print(f"[{timestamp}] WARNING:  CPU={cpu}%")
    else:
        print(f"[{timestamp}] OK:       CPU={cpu}%")

    time.sleep(POLL_INTERVAL)
```

break vs Condition-Driven Exit

Prefer a condition-driven exit (modifying the while condition itself) when the exit logic is straightforward. Prefer break when: the exit condition only becomes known partway through the loop body; there are multiple distinct exit paths; or the exit condition is a positive test ("found it") rather than a negative one ("not done yet"). Using break for every exit path in a simple loop is a code smell — it hides the loop's actual termination logic.

Example 3 — Real-world application

Binary search: break used when the target is found, condition handles the empty-range case.

```
def binary_search(sorted_list: list, target) -> int:
    """Return index of target in sorted_list, or -1 if not found."""
    low, high = 0, len(sorted_list) - 1
    result = -1

    while low <= high:       # condition: valid search space remains
        mid = (low + high) // 2
```

```
        if sorted_list[mid] == target:
            result = mid
            break            # break: found it, no need to search further
        elif sorted_list[mid] < target:
            low = mid + 1
        else:
            high = mid - 1

    return result

prices = [12.50, 18.99, 24.00, 31.75, 45.99, 52.00, 67.50]
for target in [31.75, 100.00, 12.50]:
    idx = binary_search(prices, target)
    if idx >= 0:
        print(f"Found {target} at index {idx}")
    else:
        print(f"{target} not found")
```

8.4 Loop Control: continue and pass

Two additional flow-control statements modify loop behaviour without exiting the loop entirely. continue skips the remainder of the current iteration and jumps immediately to the next condition check. pass is a syntactic placeholder that does nothing — it exists solely to satisfy Python's requirement that a block cannot be empty.

continue: Skipping Iterations

continue is useful when a loop processes a collection but some items should be skipped based on a condition. Rather than wrapping the processing logic in an if block, continue allows you to reject bad items early and keep the main processing path at minimal indentation.

Example 1 — Introduction

Processing a transaction log: skip malformed records with continue, process valid ones.

```
import csv
from io import StringIO

# Simulated CSV log with some malformed entries
raw_log = """id,amount,status
TXN-001,149.95,completed
TXN-002,,pending
TXN-003,87.50,completed
TXN-004,INVALID,completed
TXN-005,320.00,failed
"""

reader   = csv.DictReader(StringIO(raw_log))
total    = 0.0
skipped  = 0

while True:
    try:
        row = next(reader)
    except StopIteration:
```

```
        break

    if not row["amount"]:
        print(f"  Skipping {row['id']}: missing amount")
        skipped += 1
        continue

    try:
        amount = float(row["amount"])
    except ValueError:
        print(f"  Skipping {row['id']}: invalid amount {row['amount']!r}")
        skipped += 1
        continue

    if row["status"] == "completed":
        total += amount

print(f"\nCompleted total: ${total:.2f}  |  Skipped: {skipped} record(s)")
```

continue vs Nested if

Using continue to reject bad items early is generally more readable than wrapping the processing logic in a nested if. The structure "skip bad items, process the rest" reads naturally with continue; the equivalent nested if requires the reader to track indentation through the success path.

Example 2 — Variation

The same filtering logic written with a nested if (harder to follow) vs continue (cleaner).

```
readings = [22.1, None, -999, 31.4, None, 28.7, -999, 17.3]
SENTINEL  = -999

# Nested if -- success path indented inside two conditions
valid_nested = []
i = 0
while i < len(readings):
    r = readings[i]
    if r is not None:
        if r != SENTINEL:
            valid_nested.append(r)
    i += 1

# continue -- invalid cases rejected first, success path at base indent
valid_cont = []
i = 0
while i < len(readings):
    r = readings[i]
    i += 1
    if r is None:     continue
    if r == SENTINEL: continue
    valid_cont.append(r)

print(f"Valid readings: {valid_cont}")
assert valid_nested == valid_cont
```

pass: Syntactic Placeholder

pass is used when Python's syntax requires a statement but no action should be taken. In practice this arises when stubbing out a loop body during development, or when a condition branch must exist syntactically but has no current implementation. In production code, pass inside a loop body should be rare; if you find yourself using it to silence a branch, ensure the intent is clear with a comment.

```
# Stub: loop structure written before body is implemented
items_to_process = [1, 2, 3]
i = 0
while i < len(items_to_process):
    pass  # TODO: implement processing logic
    i += 1

# pass in a silent error-swallowing context (use sparingly)
import socket
ports_to_scan = [80, 443, 8080, 9090]
open_ports    = []
idx = 0
while idx < len(ports_to_scan):
    port = ports_to_scan[idx]
    idx += 1
    try:
        s = socket.create_connection(("localhost", port), timeout=0.1)
        open_ports.append(port)
        s.close()
    except OSError:
        pass    # port not open or connection refused -- intentionally ignored

print(f"Open ports: {open_ports}")
```

8.5 Flag Variables and Sentinel Values

Two patterns that appear repeatedly in while loop design are flag variables and sentinel values. Both are mechanisms for signalling state to the loop from inside the loop body.

A flag variable is a boolean that starts in one state and is set to the opposite state when a specific event occurs inside the loop. The loop condition tests the flag rather than the event condition directly. This makes the loop's purpose explicit: it runs until the flag is set.

A sentinel value is a special value that, when encountered in a data stream or user input, signals the end of input. Rather than knowing in advance how many items to process, the loop reads items until the sentinel appears.

Flag Variables

Example 1 — Introduction

A system health checker using a flag to track whether any subsystem reported a fault.

```
subsystems = [
    {"name": "API Gateway",   "status": "healthy",  "latency_ms": 45},
    {"name": "Auth Service",  "status": "degraded", "latency_ms": 920},
    {"name": "DB Primary",    "status": "healthy",  "latency_ms": 8},
    {"name": "Cache Layer",   "status": "healthy",  "latency_ms": 2},
```

```
    {"name": "Queue Worker",  "status": "down",     "latency_ms": 0},
]

fault_detected = False   # flag starts False
i = 0

while i < len(subsystems):
    sub = subsystems[i]
    i += 1

    if sub["status"] != "healthy":
        print(f"  [FAULT] {sub['name']}: {sub['status']}")
        fault_detected = True    # flag set when fault found
    else:
        print(f"  [OK]    {sub['name']}: {sub['latency_ms']}ms")

if fault_detected:
    print("\nSystem status: DEGRADED -- page on-call")
else:
    print("\nSystem status: ALL HEALTHY")
```

Sentinel Values

The sentinel pattern is standard for interactive data collection: the user enters items one at a time and types a special value (the sentinel) to signal they are done. The loop reads items until the sentinel is detected, then exits without processing the sentinel itself.

Example 2 — Variation

An expense entry tool using a blank line as the sentinel to end input.

```
print("Enter expenses (description and amount, comma-separated).")
print("Press Enter on a blank line when done.\n")

expenses = []

while True:
    raw = input("Expense: ").strip()

    if not raw:                 # blank line is the sentinel
        break

    parts = raw.split(",")
    if len(parts) != 2:
        print("  Format: description, amount")
        continue

    description = parts[0].strip()
    try:
        amount = float(parts[1].strip())
    except ValueError:
        print(f"  Invalid amount: {parts[1]!r}")
        continue

    expenses.append({"description": description, "amount": amount})
    print(f"  Added: {description} ${amount:.2f}")
```

```
if expenses:
    total = sum(e["amount"] for e in expenses)
    print(f"\n{len(expenses)} expenses totalling ${total:.2f}")
else:
    print("\nNo expenses entered.")
```

Sentinel in Data Streams

Example 3 — Real-world application

A log reader that processes lines until it encounters a specific end-of-transmission marker.

```
END_MARKER = "---END---"

# Simulated log stream (in practice this would read from a socket or file)
log_stream = [
    "2024-07-15T09:01:02 INFO  Service started",
    "2024-07-15T09:01:05 INFO  Connected to database",
    "2024-07-15T09:04:18 WARN  High memory usage: 82%",
    "2024-07-15T09:07:33 ERROR Disk write failed",
    "2024-07-15T09:07:34 INFO  Failover initiated",
    "---END---",
    "2024-07-15T09:08:00 INFO  This line is after the sentinel",
]

idx    = 0
errors = []

while idx < len(log_stream):
    line = log_stream[idx]
    idx += 1

    if line == END_MARKER:    # sentinel detected
        break

    if "ERROR" in line:
        errors.append(line)

print(f"Processed {idx - 1} line(s) before sentinel.")
print(f"Errors found: {len(errors)}")
for e in errors:
    print(f"  {e}")
```

8.6 while Loops for Data Processing

While loops are well-suited to processing data streams of unknown length, data sources that must be polled until a response is ready, and algorithms that iterate until a numerical convergence criterion is met. In all three cases the number of iterations is not known when the loop starts, which rules out a for loop over a fixed range.

Polling Until Ready

Example 1 — Introduction

Polling a job queue until a result is available, with exponential backoff.

```
import time, random

def submit_job(payload: str) -> str:
    return "JOB-7741"

def check_status(job_id: str) -> str:
    """Returns: pending | processing | complete | failed"""
    return random.choice(["pending", "pending", "processing", "complete"])

job_id      = submit_job("analyse_dataset_q3")
delay       = 1.0     # initial poll delay
MAX_DELAY   = 30.0
MAX_WAIT    = 300.0  # 5 minutes total
start       = time.time()

print(f"Job {job_id} submitted. Polling for result...")

while True:
    status = check_status(job_id)
    elapsed = time.time() - start

    if status == "complete":
        print(f"Job complete after {elapsed:.1f}s.")
        break
    elif status == "failed":
        print(f"Job failed after {elapsed:.1f}s.")
        break
    elif elapsed > MAX_WAIT:
        print(f"Timed out after {MAX_WAIT}s.")
        break

    print(f"  [{elapsed:.1f}s] Status: {status}. Retrying in {delay:.0f}s...")
    time.sleep(delay)
    delay = min(delay * 2, MAX_DELAY)    # exponential backoff
```

Numerical Convergence

Many numerical algorithms iterate until successive estimates differ by less than a target tolerance. The while loop is the natural structure: iterate while the error exceeds the threshold, and stop as soon as convergence is achieved.

Example 2 — Variation

Newton-Raphson square root algorithm: iterate until the estimate converges to a tolerance.

```
def sqrt_newton(n: float, tolerance: float = 1e-10) -> float:
    """
    Compute sqrt(n) using Newton-Raphson iteration.
    Converges quadratically: each iteration roughly doubles correct digits.
    """
    if n < 0:
        raise ValueError("Cannot compute sqrt of negative number")
```

```
    if n == 0:
        return 0.0

    estimate = n / 2.0   # initial guess
    iterations = 0

    while True:
        improved  = (estimate + n / estimate) / 2.0
        iterations += 1
        if abs(improved - estimate) < tolerance:
            break
        estimate = improved

    return improved, iterations

import math
for value in [2, 9, 100, 0.5, 1e8]:
    result, iters = sqrt_newton(value)
    error = abs(result - math.sqrt(value))
    print(f"sqrt({value:>10g}) = {result:.10f}  (iters={iters},
err={error:.2e})")
```

Sliding Window Processing

Example 3 — Real-world application

A sliding-window moving average calculator for time-series sensor data.

```
from collections import deque

readings = [22.1, 21.8, 23.4, 35.2, 22.9, 23.1, 22.7, 24.0,
            23.8, 22.5, 21.9, 22.3]

WINDOW     = 3
SPIKE_MULT = 1.5    # flag if reading > 1.5x window average

window   = deque(maxlen=WINDOW)
idx      = 0

while idx < len(readings):
    temp = readings[idx]
    idx += 1

    window.append(temp)
    if len(window) < WINDOW:
        print(f"  Reading {idx}: {temp:.1f}C  (warming up window)")
        continue

    avg = sum(window) / len(window)
    is_spike = temp > avg * SPIKE_MULT
    flag = " *** SPIKE ***" if is_spike else ""
    print(f"  Reading {idx}: {temp:.1f}C  avg={avg:.1f}C{flag}")
```

8.7 The while-else Clause

Python's while loop has an optional else clause that executes when the loop exits by the condition becoming False — but not when it exits via break. This distinction makes while-else useful for expressing the "searched and not found" or "exhausted without success" pattern: the else branch fires only when the loop ran to completion without being interrupted.

This is one of Python's less well-known features, but it is genuinely useful for linear search, retry exhaustion, and any situation where you need to distinguish between "loop ended naturally" and "loop was terminated early".

Search Without Found

Example 1 — Introduction

Finding the first available slot in a resource pool using while-else to detect exhaustion.

```
connection_pool = [
    {"id": "conn-1", "in_use": True},
    {"id": "conn-2", "in_use": True},
    {"id": "conn-3", "in_use": False},
    {"id": "conn-4", "in_use": True},
]

idx = 0
while idx < len(connection_pool):
    conn = connection_pool[idx]
    idx += 1
    if not conn["in_use"]:",
        conn["in_use"] = True
        print(f"Acquired: {conn['id']}")
        break
else:
    # Runs only when the loop ended without a break
    print("No available connections in pool. Request queued.")
```

Retry Exhaustion

Example 2 — Variation

A retry loop that uses else to cleanly distinguish success from exhaustion.

```
import random

MAX_RETRIES = 4

def unreliable_api_call() -> dict:
    if random.random() < 0.7:
        raise ConnectionError("Service temporarily unavailable")
    return {"status": "ok", "data": [1, 2, 3]}

attempt = 0
while attempt < MAX_RETRIES:
    attempt += 1
    try:
        result = unreliable_api_call()
        print(f"Success on attempt {attempt}: {result}")
```

```
            break
        except ConnectionError as e:
            print(f"Attempt {attempt} failed: {e}")
else:
    # Only reached if all MAX_RETRIES attempts failed (no break)
    print(f"All {MAX_RETRIES} attempts exhausted. Alerting on-call.")
```

while-else for Input Validation

Example 3 — Real-world application

Collecting a valid configuration value with a maximum number of attempts before aborting.

```
MAX_ATTEMPTS = 3
VALID_ENVS   = {"dev", "staging", "prod"}

attempt = 0
while attempt < MAX_ATTEMPTS:
    attempt += 1
    env = input(f"Environment ({"/".join(sorted(VALID_ENVS))}):
").strip().lower()
    if env in VALID_ENVS:
        print(f"Deploying to {env}...")
        break
    remaining = MAX_ATTEMPTS - attempt
    if remaining > 0:
        print(f"  Invalid. {remaining} attempt(s) remaining.")
else:
    # Loop exhausted without a valid entry
    print("Too many invalid attempts. Deployment aborted.")
```

Chapter 8 Summary

- A while loop evaluates its condition before each iteration and runs the body while the condition is True. It exits when the condition becomes False or a break is encountered. Every while loop needs initialisation, a condition, a body that does useful work, and an update that eventually makes the condition False.
- Compound conditions in the while header use and (all must be True to continue) and or (any True continues). and is the "run until something fails or a limit is reached" pattern. Complex multi-condition exit logic keeps the stopping rules visible in the loop header.
- while True: creates an intentional infinite loop, exited only via break, return, or an uncaught exception. It is the standard pattern for interactive menus, event listeners, server request handlers, and monitoring loops. Use it when the exit condition is most naturally expressed as a positive test inside the body.
- continue skips the remainder of the current iteration and jumps to the next condition check. It is the idiomatic way to reject bad items early in a processing loop, keeping the success path at minimal indentation. pass is a syntactic placeholder for empty blocks and is rarely needed in production code.
- A flag variable is a boolean set to signal an event; the loop condition tests the flag. A sentinel value is a special input value that signals end of input; the loop breaks when the sentinel is

encountered without processing it. Both patterns communicate intent clearly and keep the exit logic explicit.

- while loops are well-suited for data processing when the iteration count is not known in advance: polling until a condition is met, iterating until numerical convergence, and sliding-window processing of streaming data. Exponential backoff is the standard pattern for polling retry loops.
- The while-else clause executes its else block only when the loop exits by the condition becoming False, not via break. It is the idiomatic way to express "searched and not found" or "all retries exhausted", cleanly distinguishing natural loop completion from early termination.

Guided Exercises

Complete the following exercises in VS Code, IDLE, or your preferred Python environment. Each exercise targets a specific concept from this chapter.

1. Implement a connection retry manager. Write a function attempt_operation() that simulates a success or failure (use random). Write a retry loop that calls it up to MAX_RETRIES times (define as a constant), waiting RETRY_DELAY seconds between attempts (also a constant). On success, print how many attempts it took. On exhaustion, print a failure message. Use while-else to distinguish the two outcomes. Test by running the loop multiple times.

2. Build a command-line expense tracker using the sentinel pattern. The user enters lines in the format "description, amount" and presses Enter on a blank line to finish. Use continue to skip and report any lines that are malformed or have non-numeric amounts. After the sentinel, print a summary: number of valid entries, total, largest expense, and smallest expense.

3. Write a sliding-window moving average function using a while loop and a deque. Accept a list of sensor readings and a window size. For each position where a full window is available, compute the mean and flag any reading that exceeds the window mean by more than a configurable multiplier. Print a formatted table showing the reading index, value, window mean, and spike flag. Test with a realistic dataset that includes at least two clear spikes.

4. Implement a circuit breaker using a while loop. Simulate calls to an unreliable service (random success/failure rate). The circuit opens (stops accepting calls) when error_count reaches MAX_ERRORS within a TIME_WINDOW. While closed, process calls, tracking success and failure counts. When tripped, print the circuit state and stop. Use a flag variable for the tripped state and a compound while condition to enforce both the error limit and the time window.

5. Write a Newton-Raphson cube root solver (cbrt(n) = n ** (1/3)) using a while loop. Start with an initial estimate of n/3, and iterate using the update rule: new = (2*estimate + n/estimate**2) / 3. Loop until successive estimates differ by less than a tolerance (default 1e-10). Print the result, the number of iterations, and the error compared to n**(1/3). Test with both positive and negative inputs and very large and very small values.

Challenge Tasks

The following tasks are open-ended. No hints are provided. They require you to design a solution, not just apply a pattern.

1. Design and implement a job queue processor. Maintain a queue (list) of jobs, each represented as a dictionary with an id, priority (1-5), payload, and status. Write a while loop that: dequeues the highest-priority pending job, simulates processing it (random success/failure), updates its status, handles transient failures by re-queuing with an incremented attempt count (up to MAX_ATTEMPTS), and logs each action. The loop should run until either the queue is empty or a configurable MAX_RUNTIME_SECONDS has elapsed. Use while-else to report whether the queue was fully drained or time expired.

2. Implement a bisection root-finder for any continuous function f(x). The algorithm: given an interval [a, b] where f(a) and f(b) have opposite signs, repeatedly halve the interval by computing the midpoint m = (a+b)/2, evaluating f(m), and replacing either a or b with m depending on sign. Loop until the interval width is less than a tolerance. Apply it to find: (1) the x where x^3 - x - 2 = 0 in [1, 2]; (2) the x where cos(x) = x in [0, 1]; (3) the break-even quantity where revenue = cost in a cost-revenue model of your design. Report the root, number of iterations, and error estimate for each.

3. Build a complete interactive command-line tool for managing a contact directory. Use a while True: menu loop as the outer structure. Implement at least six commands: add, search (by name or email), update, delete, list all, and export to CSV. Each command's logic should be in its own function. Input collection within commands should use validation loops (while True with continue) for each field, enforcing appropriate constraints (non-empty name, valid email format containing @, phone digits only). The export function should write a properly formatted CSV file. Demonstrate the tool handles invalid inputs at every point without crashing.

CHAPTER 9 **Looping with for**

Think, Code, Create – A Python Programming Journey

The for loop iterates over the elements of any iterable object — list, tuple, string, dictionary, file, generator, or any object that implements the iteration protocol. It is the natural choice whenever the collection to iterate over is known before the loop starts. Unlike while, which runs until a condition changes, for guarantees exactly one iteration per element with no possibility of accidentally running forever.

This chapter covers the full for loop toolkit: basic iteration over common types, range() for index-based and count-driven loops, nested loops for multi-level data, zip() and enumerate() for parallel and indexed iteration, list comprehensions as the Pythonic shorthand for the filter-and-transform pattern, common looping idioms, and break/continue/for-else for flow control. Real-world examples throughout replace abstract counting exercises.

9.1 Introduction to for Loops

A for loop binds a loop variable to each element of an iterable in turn and executes the loop body once per element. When the iterable is exhausted, the loop exits. The loop variable takes the value of the current element at the start of each iteration; it can be any valid Python name. For compound elements (tuples, for example), multiple names can be used to unpack the element directly in the for statement.

Iterating Over Common Types

Example 1 — Introduction

for loop over a list of product records, a dictionary, and a string — the three most common iterables.

```
# Iterating a list of dictionaries
products = [
    {"sku": "A-100", "name": "Widget",    "price": 12.50, "stock": 45},
    {"sku": "B-200", "name": "Gadget",    "price": 89.99, "stock": 0},
    {"sku": "C-300", "name": "Doohickey","price": 5.00,   "stock": 200},
]

for product in products:
    status = "IN STOCK" if product["stock"] > 0 else "OUT OF STOCK"
    print(f"  {product['sku']}  {product['name']:<12}
${product['price']:>7.2f}  {status}")

# Iterating dictionary items (key-value pairs)
config = {"host": "db-prod-01", "port": 5432, "timeout": 30, "pool_size": 10}
for key, value in config.items():
    print(f"  {key:<12} = {value}")

# Iterating a string character by character
```

```
api_key = "AK-7741-PROD"
digit_count = sum(1 for ch in api_key if ch.isdigit())
print(f"API key contains {digit_count} digits")
```

Unpacking in the for Statement

When each element of the iterable is itself a sequence (a tuple, list, or any iterable), multiple variable names in the for statement unpack that element directly. This eliminates the need to index into the element inside the loop body and produces cleaner code.

Example 2 — Variation

Tuple unpacking in a for loop: exchange rates, coordinate pairs, and named record fields.

```
# Two-element tuples: unpack to two variables
exchange_rates = [("AUD", 1.52), ("EUR", 1.07), ("GBP", 0.81), ("JPY", 157.4)]
amount_usd = 1_000

for currency, rate in exchange_rates:
    converted = amount_usd * rate
    print(f"  USD {amount_usd:,} = {currency} {converted:,.2f}")

# Three-element tuples: server connection parameters
servers = [
    ("db-primary",   "10.0.1.10", 5432),
    ("db-replica-1", "10.0.1.11", 5432),
    ("cache",        "10.0.1.20", 6379),
]

for name, host, port in servers:
    print(f"  Checking {name} ({host}:{port})...")
```

for-else

Like while, the for loop supports an optional else clause. The else block executes when the loop exhausts the iterable normally — but not when it exits via break. This is the idiomatic pattern for "search and report not found": iterate, break if found, else report failure.

Example 3 — Real-world application

for-else to find a first available worker, reporting pool exhaustion if none found.

```
workers = [
    {"id": "W-01", "active": True,  "load": 95},
    {"id": "W-02", "active": False, "load": 0},
    {"id": "W-03", "active": True,  "load": 88},
    {"id": "W-04", "active": True,  "load": 72},
]

LOAD_LIMIT = 80

for worker in workers:
    if worker["active"] and worker["load"] < LOAD_LIMIT:
        print(f"Assigned to {worker['id']} (load: {worker['load']}%)")
        break
else:
```

```
    print("No available worker. Job queued.")
```

9.2 Iterating Through a Range with for

range(stop), range(start, stop), and range(start, stop, step) generate integer sequences without creating a list in memory. range() is lazy: it produces the next value only when requested, making it memory-efficient regardless of the range size. range objects are immutable and support len(), membership testing with in, and slicing.

The most common use of range() with for is count-driven iteration: when you need to repeat something a fixed number of times or generate indices into a list. When you need both an index and a value, prefer enumerate() (section 9.6). When you just need to iterate over the collection, iterate directly without range(len(...)).

Count-Driven Iteration

Example 1 — Introduction

range() for batch processing: split a large dataset into fixed-size chunks for parallel upload.

```
records = list(range(1, 10_001))   # simulated dataset of 10,000 records
BATCH_SIZE = 500

total_batches = (len(records) + BATCH_SIZE - 1) // BATCH_SIZE   # ceiling
division

for batch_num in range(total_batches):
    start = batch_num * BATCH_SIZE
    end   = min(start + BATCH_SIZE, len(records))
    batch = records[start:end]
    print(f"Batch {batch_num + 1:>3}/{total_batches}: records {start+1:>5}-
{end:>5} ({len(batch)} records)")
```

Step Values and Countdown

Example 2 — Variation

Positive and negative step values: processing every Nth item and a countdown sequence.

```
# Positive step: sample every 5th reading from a sensor log
sensor_log = [round(20 + i * 0.3 + (-1)**i * 0.5, 1) for i in range(50)]
SAMPLE_INTERVAL = 5

print("Sampled readings:")
for i in range(0, len(sensor_log), SAMPLE_INTERVAL):
    print(f"  Index {i:>3}: {sensor_log[i]:.1f}C")

# Negative step: countdown to a deployment window
COUNTDOWN_FROM = 10
print("\nDeployment countdown:")
for t in range(COUNTDOWN_FROM, 0, -1):
    print(f"  T-{t:>2}s: systems nominal")
print("  DEPLOY")
```

Avoiding range(len(...))

range(len(collection)) to iterate a collection is an anti-pattern in Python. It is more verbose than direct iteration and requires manual indexing. The idiomatic alternative is to iterate the collection directly (for item in collection:) or use enumerate() when the index is also needed. Reserve range() for when you genuinely need a sequence of integers, not as a device to access a list.

Example 3 — Real-world application

Comparing the anti-pattern, the idiomatic replacement, and the case where range() is genuinely needed.

```
transactions = [149.95, 87.50, 320.00, 45.99, 1_200.00]

# Anti-pattern: range(len()) just to access the list
for i in range(len(transactions)):
    print(transactions[i])          # unnecessary indexing

# Idiomatic: iterate directly
for amount in transactions:
    print(amount)

# When range() is genuinely needed: generating indices for a sparse update
# (updating every other record in-place requires the index)
for i in range(0, len(transactions), 2):
    transactions[i] = round(transactions[i] * 1.05, 2)  # 5% surcharge
print("After surcharge:", transactions)
```

9.3 Nested for Loops

A nested for loop places one loop inside the body of another. For each iteration of the outer loop, the inner loop runs completely. This is the natural structure for processing multi-level data: rows and columns of a table, pairs drawn from two collections, items within categories, or sub-items within items.

The total iteration count is the product of the two loop sizes. A loop over 100 rows, each with 50 columns, performs 5,000 iterations. This compounding grows quickly with depth and data size, so nested loops on large datasets warrant consideration of more efficient approaches — list comprehensions, vectorised operations, or algorithmic improvements.

Two-Level Data Structures

Example 1 — Introduction

Processing a table of monthly regional sales: outer loop over regions, inner loop over months.

```
sales_data = {
    "NSW": [142_000, 138_500, 155_000, 148_200, 162_000, 171_000],
    "VIC": [128_000, 131_200, 129_800, 141_500, 138_900, 152_000],
    "QLD": [98_000,  102_400, 110_200, 108_700, 115_000, 119_400],
}
MONTHS = ["Jan", "Feb", "Mar", "Apr", "May", "Jun"]

  print("Region".ljust(6) + "  " + "  ".join(m.rjust(8) for m in MONTHS) + "
Total")
print("-" * 72)
```

```
for region, monthly in sales_data.items():
    total = 0
    row   = f"{region:<6}  "
    for amount in monthly:
        row   += f"${amount/1000:>6.1f}K  "
        total += amount
    print(row + f"${total/1000:>8.1f}K")
```

Cross-Product Iteration

When every combination of elements from two collections must be considered, nested loops produce the Cartesian product. This pattern appears in pricing matrices, feature combination testing, and constraint checking.

Example 2 — Variation

Generating a freight rate matrix: every combination of weight tier and destination zone.

```
WEIGHT_TIERS  = [("0-5 kg", 1.0), ("5-20 kg", 1.8), ("20-50 kg", 3.2), (">50
kg", 5.5)]
DEST_ZONES    = [("Metro",      8.50), ("Regional", 14.20),
                 ("Remote",    22.00), ("Int'l",    45.00)]

# Header
  header = "Weight".ljust(12) + "".join(z[0].rjust(12) for z in DEST_ZONES)
print(header)
print("-" * (12 + 12 * len(DEST_ZONES)))

for tier_label, weight_mult in WEIGHT_TIERS:
    row = f"{tier_label:<12}"
    for zone_label, base_rate in DEST_ZONES:
        rate = base_rate * weight_mult
        row += f"${rate:>10.2f}  "
    print(row)
```

Nested Loops with break and continue

break inside a nested loop exits only the innermost loop. To break from an outer loop based on something found in an inner loop, either set a flag variable or restructure the logic into a function and use return.

Example 3 — Real-world application

Searching a matrix for the first cell exceeding a threshold: inner break, outer flag.

```
# Grid of sensor readings (rows=time, cols=sensor_id)
readings = [
    [12.1, 14.3, 11.8, 13.2],
    [13.4, 15.1, 18.7, 12.9],
    [14.2, 16.8, 45.1, 13.7],    # sensor[2] has a spike at time[2]
    [13.8, 15.2, 17.4, 14.1],
]
SPIKE_THRESHOLD = 40.0

found = False
```

```
for time_idx, row in enumerate(readings):
    for sensor_idx, value in enumerate(row):
        if value > SPIKE_THRESHOLD:
            print(f"SPIKE: t={time_idx} sensor={sensor_idx} value={value}")
            found = True
            break       # exits inner loop
    if found:
        break           # exits outer loop

if not found:
    print("No spikes detected.")
```

9.4 Iterating with zip() and enumerate()

Two built-in functions dramatically improve the readability of loops that process multiple sequences simultaneously or need access to the current element's position: zip() pairs up corresponding elements from multiple iterables, and enumerate() adds an index counter to any iterable. Both return lazy iterators: they generate values on demand rather than creating a full list in memory.

zip(): Parallel Iteration

zip(a, b, c, ...) yields tuples containing one element from each argument, stopping when the shortest is exhausted. itertools.zip_longest() continues to the end of the longest, filling missing positions with a configurable fill value (default None). Use zip() to process related parallel sequences together, eliminating the need for index-based access.

Example 1 — Introduction

zip() pairing product records with live inventory counts from two separate data sources.

```
from itertools import zip_longest

# Two parallel lists from different API endpoints
catalogue = [
    {"sku": "P-001", "name": "Ethernet Cable", "price": 12.50},
    {"sku": "P-002", "name": "USB Hub",         "price": 34.99},
    {"sku": "P-003", "name": "HDMI Adapter",    "price": 22.00},
    {"sku": "P-004", "name": "Webcam",          "price": 89.95},
]
live_stock = [45, 0, 12, 3]   # from warehouse system (same order)

  print(f"  SKU     Product              Price   Stock  Status")
print("-" * 60)

for product, stock in zip(catalogue, live_stock):
    status = "OK" if stock > 5 else ("LOW" if stock > 0 else "OUT")
    print(f"{product['sku']:<8} {product['name']:<20}"
          f" ${product['price']:>7.2f} {stock:>6}   {status}")
```

zip() for Pairwise Comparison

Example 2 — Variation

zip() pairing consecutive elements to compute period-over-period changes in a time series.

```
monthly_revenue = [42_000, 38_500, 51_000, 47_800, 55_200, 61_000]
MONTHS = ["Jan", "Feb", "Mar", "Apr", "May", "Jun"]

# zip current month with previous month to compute month-on-month change
for (prev_m, prev_r), (curr_m, curr_r) in zip(
        zip(MONTHS, monthly_revenue),
        zip(MONTHS[1:], monthly_revenue[1:])):",
    change     = curr_r - prev_r
    change_pct = change / prev_r * 100
    arrow = "▲" if change >= 0 else "▼"
    print(f"  {prev_m}→{curr_m}: ${curr_r:>8,}
{arrow}{abs(change_pct):.1f}%")
```

enumerate(): Index and Value Together

enumerate(iterable, start=0) yields (index, value) tuples. It replaces the pattern of manually maintaining an index counter or using range(len(collection)), both of which are considered non-idiomatic in Python. enumerate() is the correct tool whenever the loop needs both the position of an element and the element itself.

Example 3 — Real-world application

enumerate() for ranked output, selective update by position, and finding a first match.

```
# Ranked leaderboard: start=1 for human-readable numbering
scores = [
    ("Chen, Wei",       9_840),
    ("Okafor, Amara",   9_210),
    ("Silva, Marcus",   8_990),
    ("Nguyen, Linh",    8_750),
]
MEDALS = {1: "U0001f947", 2: "U0001f948", 3: "U0001f949"}

for rank, (name, score) in enumerate(scores, start=1):
    medal = MEDALS.get(rank, "  ")
    print(f"  {medal} #{rank}  {name:<22} {score:>6,} pts")

# Find first record failing a validation rule
records = [
    {"id": "R-01", "amount": 149.95},
    {"id": "R-02", "amount": -10.00},
    {"id": "R-03", "amount":  87.50},
]
for i, record in enumerate(records):
    if record["amount"] < 0:
        print(f"\nFirst invalid record at index {i}: {record}")
        break
```

9.5 List Comprehensions

A list comprehension is a concise expression that creates a new list by applying an expression to each element of an iterable, optionally filtering elements based on a condition. It is the idiomatic Python replacement for the "initialise empty list, loop, append" pattern, and is generally faster because the operation is implemented in C internally.

Syntax: [expression for variable in iterable] or [expression for variable in iterable if condition]. Dict and set comprehensions follow the same structure with {} instead of []; generator expressions use () and produce values lazily without creating a list.

Filter and Transform

Example 1 — Introduction

List comprehensions replacing explicit append loops for filtering and transforming product data.

```
products = [
    {"sku": "A-100", "price": 12.50, "stock": 45,  "active": True},
    {"sku": "B-200", "price": 89.99, "stock": 0,   "active": True},
    {"sku": "C-300", "price": 5.00,  "stock": 200, "active": False},
    {"sku": "D-400", "price": 34.95, "stock": 12,  "active": True},
]

# Filter: active products in stock
available = [p for p in products if p["active"] and p["stock"] > 0]
print(f"Available: {[p['sku'] for p in available]}")

# Transform: apply a 10% discount to all active products
discounted = [
    {**p, "price": round(p["price"] * 0.9, 2)}
    for p in products if p["active"]
]
for p in discounted:
    print(f"  {p['sku']}: ${p['price']:.2f}")
```

Nested Comprehensions

List comprehensions can be nested: [expr for x in outer for y in inner]. This is equivalent to two nested for loops and is appropriate for flattening a list-of-lists or computing Cartesian products, provided the result remains readable. When the logic becomes complex, a plain nested loop is usually clearer.

Example 2 — Variation

Flattening a nested structure and computing a Cartesian product with nested comprehensions.

```
# Flatten a list-of-lists (e.g. log lines grouped by server)
server_logs = [
    ["api-01: 200 /health", "api-01: 404 /missing"],
    ["db-01: query 12ms",   "db-01: query 8ms"],
    ["cache-01: hit",       "cache-01: miss", "cache-01: hit"],
]

all_lines = [line for server in server_logs for line in server]
print(f"Total log lines: {len(all_lines)}")
```

```
# All (size, colour) SKU variants
sizes   = ["S", "M", "L", "XL"]
colours = ["Black", "White", "Navy"]
skus    = [f"{colour[:3].upper()}-{size}" for size in sizes for colour in
colours]
print(f"Generated {len(skus)} SKUs: {skus[:6]}...")
```

Dict Comprehensions and Generator Expressions

Example 3 — Real-world application

Dict comprehension to invert a lookup table; generator expression for memory-efficient aggregation.

```
# Dict comprehension: invert a currency-code to name mapping
code_to_name = {"AUD": "Australian Dollar", "USD": "US Dollar",
                "EUR": "Euro", "GBP": "British Pound"}

name_to_code = {name: code for code, name in code_to_name.items()}
print(name_to_code["Euro"])    # EUR

# Dict comprehension: build an index by SKU for O(1) lookup
products = [{"sku": "P-001", "price": 12.50}, {"sku": "P-002", "price":
89.99}]
by_sku = {p["sku"]: p for p in products}
print(by_sku["P-002"])         # {"sku": "P-002", "price": 89.99}

# Generator expression: sum revenue without creating an intermediate list
# (efficient for large datasets -- does not materialise the full list in
memory)
amounts = [149.95, 87.50, 0, 320.00, 45.99, 1_200.00]
total   = sum(a for a in amounts if a > 0)
count   = sum(1 for a in amounts if a > 0)
print(f"Total: ${total:,.2f}  Count: {count}")
```

9.6 Iterating Over a List with Indexes

Although direct iteration (for item in collection:) is idiomatic for most cases, there are legitimate situations where the index is needed during iteration: updating elements in place, comparing adjacent elements, generating position-dependent output, and implementing algorithms that need to look ahead or behind the current element. In all these cases, enumerate() is the preferred approach over range(len(...)).

In-Place Updates and Positional Logic

Example 1 — Introduction

enumerate() for position-dependent processing: alternating surcharges and in-place correction.

```
invoices = [
    {"id": "INV-001", "amount": 1_250.00},
    {"id": "INV-002", "amount":   875.00},
    {"id": "INV-003", "amount": 3_200.00},
    {"id": "INV-004", "amount":   450.00},
```

```
    {"id": "INV-005", "amount": 2_100.00},
]

# Apply priority surcharge to every other invoice (index 0, 2, 4...)
PRIORITY_SURCHARGE = 0.05
for i, inv in enumerate(invoices):
    if i % 2 == 0:
        original = inv["amount"]
        invoices[i]["amount"] = round(original * (1 + PRIORITY_SURCHARGE), 2)
        print(f"  {inv['id']}: ${original:,.2f} + surcharge =
${invoices[i]['amount']:,.2f}")
    else:
        print(f"  {inv['id']}: ${inv['amount']:,.2f}  (standard)")
```

Adjacent Element Comparison

Comparing each element to the next requires knowing the current index. This is a case where range(len()-1) with indexing, or zip(seq, seq[1:]), are both appropriate. The zip approach is slightly more Pythonic.

Example 2 — Variation

Detecting sign changes in a time series to identify trend reversals.

```
closing_prices = [142.50, 145.20, 144.80, 148.10, 146.70,
                  149.30, 152.00, 150.40, 153.80, 151.20]

reversal_count = 0

for i, (prev, curr) in enumerate(zip(closing_prices, closing_prices[1:]),
start=1):
    if prev > closing_prices[i-1 if i > 1 else 0]:
        prev_trend = "up"
    else:
        prev_trend = "down"

# Simpler: detect direction changes
for i in range(1, len(closing_prices) - 1):
    prev_delta = closing_prices[i]   - closing_prices[i-1]
    next_delta = closing_prices[i+1] - closing_prices[i]
    if (prev_delta > 0) != (next_delta > 0):
        direction = "peak" if prev_delta > 0 else "trough"
        reversal_count += 1
        print(f"  Day {i+1}: ${closing_prices[i]:.2f}  ({direction})")

print(f"\n{reversal_count} trend reversals detected")
```

Generating Positional Output

Example 3 — Real-world application

enumerate() for numbered report output and finding the Nth occurrence in a dataset.

```
transactions = [
    {"id": "T-001", "type": "sale",   "amount": 1_249.00},
    {"id": "T-002", "type": "refund", "amount":  149.50},
```

```
    {"id": "T-003", "type": "sale",   "amount":   89.99},
    {"id": "T-004", "type": "sale",   "amount":  450.00},
    {"id": "T-005", "type": "refund", "amount":   89.99},
]

# Numbered report (1-based)
for n, txn in enumerate(transactions, start=1):
    sign = "+" if txn["type"] == "sale" else "-"
    print(f"  {n:>3}. {txn['id']}  {sign}${txn['amount']:>8,.2f}")

# Find the second refund
TARGET_TYPE = "refund"
TARGET_N    = 2
found = 0
for i, txn in enumerate(transactions):
    if txn["type"] == TARGET_TYPE:
        found += 1
        if found == TARGET_N:
            print(f"\n{TARGET_N}nd {TARGET_TYPE} at index {i}: {txn}")
            break
```

9.7 Common Looping Patterns

Several high-level patterns appear repeatedly in Python programs that process collections. Recognising them by name makes it easier to select the right tool — a for loop, a comprehension, a built-in function like sum() or max(), or a function from itertools. This section demonstrates the five most common patterns on realistic data.

Filter, Transform, Accumulate

Example 1 — Introduction

The three fundamental collection patterns applied to a sales transaction dataset.

```
transactions = [
    {"id": "T-001", "status": "completed", "amount": 1_249.00, "region":
"AU"},
    {"id": "T-002", "status": "refunded",  "amount":   149.50, "region":
"AU"},
    {"id": "T-003", "status": "completed", "amount":    89.99, "region":
"NZ"},
    {"id": "T-004", "status": "pending",   "amount":   450.00, "region":
"AU"},
    {"id": "T-005", "status": "completed", "amount": 3_200.00, "region":
"NZ"},
]

# FILTER: completed transactions only
completed = [t for t in transactions if t["status"] == "completed"]

# TRANSFORM: extract amounts from completed transactions
amounts = [t["amount"] for t in completed]

# ACCUMULATE: sum, min, max using built-ins (preferred over explicit loops)
print(f"Completed:  {len(completed)} transactions")
print(f"Total:      ${sum(amounts):>10,.2f}")
```

```
print(f"Average:    ${sum(amounts)/len(amounts):>10,.2f}")
print(f"Largest:    ${max(amounts):>10,.2f}")
print(f"Smallest:   ${min(amounts):>10,.2f}")
```

Group By

Grouping elements by a key is a frequent task: totalling sales by region, counting errors by type, aggregating metrics by time period. The standard pattern uses a dict to accumulate values under each key.

Example 2 — Variation

Grouping completed transactions by region to produce a regional revenue summary.

```
from collections import defaultdict

region_totals  = defaultdict(float)
region_counts  = defaultdict(int)

for txn in transactions:
    if txn["status"] == "completed":
        region_totals[txn["region"]] += txn["amount"]
        region_counts[txn["region"]] += 1

  print(f"  Region   Count        Total      Average")
print("-" * 42)
for region in sorted(region_totals):
    count = region_counts[region]
    total = region_totals[region]
    print(f"{region:<8} {count:>6} ${total:>11,.2f} ${total/count:>11,.2f}")
```

Pair-wise and Windowed Iteration

Example 3 — Real-world application

Pair-wise comparison using zip(seq, seq[1:]) and a sliding window using itertools.islice.

```
from itertools import islice
from collections import deque

cpu_readings = [42, 45, 48, 71, 85, 92, 89, 78, 65, 55, 48, 44]

# Pair-wise: detect when CPU increases by more than 20% in one interval
SPIKE_JUMP = 20
print("Sudden CPU increases:")
for prev, curr in zip(cpu_readings, cpu_readings[1:]):
    if curr - prev >= SPIKE_JUMP:
        print(f"  {prev}% -> {curr}%  (+{curr-prev}%)")

# Sliding window average (window=3)
WINDOW = 3
window = deque(maxlen=WINDOW)
averages = []

for reading in cpu_readings:
```

```
    window.append(reading)
    if len(window) == WINDOW:
        averages.append(round(sum(window) / WINDOW, 1))

print(f"\nSmoothed (window={WINDOW}): {averages}")
```

Chapter 9 Summary

- The for loop iterates over any iterable — list, tuple, dict, string, range, generator, or any object implementing the iteration protocol. Compound elements can be unpacked directly in the for statement. The for-else clause executes only when the loop exhausts the iterable without a break, making it the idiomatic "search and not found" pattern.
- range(stop), range(start, stop), range(start, stop, step) generate lazy integer sequences. Use range() for count-driven repetition and generating indices. Avoid range(len(collection)) for direct iteration — iterate the collection directly or use enumerate() when the index is needed.
- Nested for loops process multi-level data. Total iterations = outer count × inner count, which can grow quickly. break exits only the innermost loop; use a flag variable or function-with-return to exit from an outer loop based on an inner condition.
- zip(a, b, ...) yields tuples of corresponding elements from parallel iterables, stopping at the shortest. itertools.zip_longest() continues to the end of the longest, filling gaps. Use zip() to process related parallel sequences; use zip(seq, seq[1:]) for pairwise adjacent-element comparisons.
- List comprehensions [expr for x in iterable if condition] replace the append-loop pattern for filter and transform operations and are generally faster. Dict comprehensions {k: v for ...} build dictionaries in one expression. Generator expressions (expr for ...) produce values lazily, using minimal memory regardless of size.
- enumerate(iterable, start=0) yields (index, value) pairs, eliminating manual counter maintenance and range(len()). Use start=1 for human-readable 1-based numbering. Combine with zip() for indexed parallel iteration.
- Five core patterns: Filter (select elements meeting a condition), Transform (apply a function to every element), Accumulate (build a running total with sum(), max(), min()), Group-by (partition into categories using a defaultdict), and Pair-wise/windowed (compare adjacent elements with zip(seq, seq[1:]) or a sliding window).

Guided Exercises

Complete the following exercises in VS Code, IDLE, or your preferred Python environment. Each exercise targets a specific concept from this chapter.

1. Write a batch processor using range(). Given a list of at least 20 records (define them as a list of dicts), split them into batches of a configurable BATCH_SIZE using range() to generate the slice indices. For each batch, print the batch number, the range of record indices it covers, the count of records, and a summary statistic (e.g. mean amount). Use ceiling division to handle the final partial batch correctly.

2. Build a regional sales report using zip() and enumerate(). You have two parallel lists: one of region names and one of monthly revenue figures (each a list of 6 months). Use zip() to pair them, then for each region compute the total and month-on-month changes using zip(months, months[1:]). Print a formatted table with region name, monthly figures, total, and the single largest month-on-month change.

3. Write a product catalogue processor using list and dict comprehensions. Given a list of product dicts (SKU, name, price, stock, active, category), produce: (1) a list of in-stock active products under $50, (2) a dict mapping SKU to discounted price (10% off all active products), (3) a dict grouping products by category, and (4) a generator expression that computes total inventory value without creating an intermediate list. Print the results of each.

4. Implement a time-series anomaly detector using nested for loops and the for-else clause. Given a 2D grid of sensor readings (rows=time steps, cols=sensor IDs), find the first cell where the reading exceeds a SPIKE_THRESHOLD using nested loops and a flag variable to break both. Also report the percentage of readings above a WARN_THRESHOLD using a list comprehension. Print a formatted summary.

5. Write a leaderboard generator using enumerate() and zip(). Given a list of competitor names and their scores in three events (three parallel lists), compute each competitor's total score, rank them using sorted(), and print a formatted leaderboard using enumerate(start=1) with gold/silver/bronze medals for the top three. Also print pairwise score differences between adjacent ranks using zip().

Challenge Tasks

The following tasks are open-ended. No hints are provided. They require you to design a solution, not just apply a pattern.

1. Build a data pipeline processor. Given a list of raw transaction records (dicts with fields including date, amount, currency, status, region, and customer_id), write a pipeline of transformations using a combination of list comprehensions, filter expressions, and for loops: (1) filter to completed transactions in the last 30 days, (2) convert all amounts to AUD using a rate table, (3) group by region and compute count, total, and average per region, (4) identify the top-3 customers by total spend, (5) flag any customer with more than 3 transactions in the period. Print a complete summary report.

2. Implement a matrix operations library using nested for loops and comprehensions. Without using NumPy, write functions for: matrix addition (element-wise), matrix multiplication (standard dot-product), matrix transposition, and finding the minimum and maximum values with their (row, col) positions. Test each function with at least three test cases including edge cases (non-square matrices, 1x1, single-row). Verify your matrix multiplication result for at least one case by hand.

3. Design and implement a configurable report generator. Given a dataset of records (at least 20 dicts with at least five fields), and a report specification dict that defines: which fields to include, any filters (field, operator, value), a group-by field, sort order (field and direction), and up to two aggregation functions (sum, mean, count, min, max) per numeric field, generate the report dynamically. The specification should be data (not code), and the report generator should interpret it using for loops and comprehensions. Demonstrate with at least three different report specifications on the same dataset.

CHAPTER 10 Working with Strings

Think, Code, Create – A Python Programming Journey

> Strings are the primary data type for text in Python, and text is everywhere in software: log files, API responses, configuration values, user input, file paths, identifiers, report output. Knowing how to index, slice, search, split, join, format, and validate strings is not optional — it is the skill that makes the difference between programs that process real data and programs that only handle perfectly clean input.
>
> This chapter covers strings systematically: their nature as immutable sequences, their rich method library, formatting for precise tabular output, splitting and parsing structured text, searching and pattern checking, comparison and normalisation, and a final section of practical text-processing patterns drawn from log parsing, data cleaning, and identifier manipulation.

10.1 Strings as Sequences

A string is an immutable ordered sequence of Unicode characters. "Immutable" means no individual character can be changed after the string is created — operations that appear to modify a string always produce a new one. "Sequence" means strings support indexing, slicing, iteration, len(), membership testing with in, and concatenation with +, just like lists.

Python string literals can use single quotes, double quotes, or triple quotes (for multi-line strings or strings that contain both quote types). Raw strings (prefixed with r) suppress backslash interpretation, making them essential for file paths and regular expression patterns.

Indexing and Slicing

Example 1 — Introduction

Indexing and slicing applied to parsing a structured log entry string.

```
# Structured log line: "2024-07-15T09:14:02 ERROR  disk quota exceeded"
log_entry = "2024-07-15T09:14:02 ERROR  disk quota exceeded"

# Indexing: individual characters
print(log_entry[0])          # "2"  -- first character
print(log_entry[-1])         # "d"  -- last character

# Slicing: [start:stop:step]
date_part = log_entry[0:10]         # "2024-07-15"
time_part = log_entry[11:19]        # "09:14:02"
level     = log_entry[20:25].strip() # "ERROR"
message   = log_entry[27:]          # "disk quota exceeded"

print(f"Date:    {date_part}")
print(f"Time:    {time_part}")
print(f"Level:   {level}")
```

```python
print(f"Message: {message}")

# Reverse a string with step -1
api_key = "AK-2024-PROD"
print(api_key[::-1])    # "DORP-4202-KA"
```

Immutability and Raw Strings

Example 2 — Variation

Demonstrating string immutability, raw strings for paths, and multi-line strings for templates.

```python
# Immutability: strings cannot be modified in place
server_name = "api-prod-01"
# server_name[4] = "X"   # TypeError: str does not support item assignment

# To "change" a string, create a new one
server_name = server_name.replace("prod", "staging")
print(server_name)  # "api-staging-01"

# Raw strings: backslashes are literal (no escape processing)
windows_path = r"C:\Users\admin\AppData\Local\Logs"
regex_pattern = r"\d{4}-\d{2}-\d{2}"   # date pattern, no extra escaping
print(windows_path)

# Triple-quoted multi-line string (preserves newlines and indentation)
email_template = """Dear {name},

Your order {order_id} has been dispatched.
Expected delivery: {delivery_date}.

Thank you."""
print(email_template.format(name="Taylor", order_id="ORD-9921",
delivery_date="18 Jul"))
```

String Membership and Iteration

Example 3 — Real-world application

Iterating over a string character by character and using in for membership testing in a data validator.

```python
def validate_identifier(identifier: str) -> list[str]:
    """Validate a service identifier: alphanumeric and hyphens only, 4-32
chars."""
    errors = []
    ALLOWED = set("abcdefghijklmnopqrstuvwxyz0123456789-")

    if not (4 <= len(identifier) <= 32):
        errors.append(f"Length must be 4-32 chars (got {len(identifier)})")

    for i, ch in enumerate(identifier):
        if ch not in ALLOWED:
            errors.append(f"Invalid character {ch!r} at position {i}")
            break    # report only the first bad character
```

```
    if identifier.startswith("-") or identifier.endswith("-"):
        errors.append("Cannot start or end with a hyphen")

    return errors

for test in ["api-gateway", "DB_PRIMARY", "x", "valid-service-name-01", "ok"]:
    issues = validate_identifier(test)
    status = "OK" if not issues else f"FAIL: {issues[0]}"
    print(f"  {test!r:<30} {status}")
```

10.2 String Methods

Python strings have a rich method library — over 40 built-in methods covering case conversion, whitespace handling, search, replacement, validation, alignment, and more. Methods never modify the original string (because strings are immutable); they always return a new string. Calls can be chained: s.strip().lower().replace("-", "_") applies three operations in sequence.

Case Conversion and Normalisation

lower() and upper() are standard for case normalisation. casefold() is more aggressive than lower() — it handles characters like the German ß (which lower() leaves as ß but casefold() converts to "ss") and is the correct method for case-insensitive comparison of text that may contain non-ASCII characters. title() capitalises the first letter of each word; capitalize() capitalises only the first letter of the string.

Example 1 — Introduction

Case normalisation for deduplication, case-insensitive lookup, and display formatting.

```
# casefold() vs lower() for robust case-insensitive comparison
entries  = ["Sydney", "SYDNEY", "sydney", "Südney"]
seen     = set()
unique   = []
for e in entries:
    key = e.casefold()
    if key not in seen:
        seen.add(key)
        unique.append(e)
print(f"Unique entries: {unique}")

# Case-insensitive lookup in a command dispatcher
COMMANDS = {"start", "stop", "restart", "status", "help"}
user_input = "  RESTART  "
cmd = user_input.strip().lower()

if cmd in COMMANDS:
    print(f"Executing: {cmd}")
else:
    print(f"Unknown command. Valid: {sorted(COMMANDS)}")

# title() for display formatting of stored-lowercase data
raw_name = "sarah jane SMITH"
display  = raw_name.title()
print(f"Display name: {display}")
```

Stripping, Replacing, and Counting

strip() removes leading and trailing characters (whitespace by default, or a specified character set). lstrip() and rstrip() strip from one side only. replace(old, new, count=-1) replaces occurrences of old with new; the optional count limits replacements. count(sub) counts non-overlapping occurrences of a substring.

Example 2 — Variation

Cleaning and normalising raw CSV field values: stripping, replacing, and counting.

```
raw_fields = [
    "  Widget A  ",
    "ACME Corp.",
    "N/A",
    "$ 1,249.00",
    "email@example.com  ",
]

for field in raw_fields:
    cleaned = field.strip()

    # Normalise price strings: remove $ and commas, then convert
    if cleaned.startswith("$"):
        price_str = cleaned.replace("$", "").replace(",", "").strip()
        try:
            price = float(price_str)
            print(f"  Price: ${price:,.2f}")
        except ValueError:
            print(f"  Invalid price: {cleaned!r}")
    elif cleaned == "N/A":
        print(f"  Missing value (substituting None)")
    else:
        print(f"  Field: {cleaned!r}")

# count() for quick frequency analysis
log_line = "ERROR: connection refused ERROR: timeout ERROR: disk full"
error_count = log_line.count("ERROR")
print(f"\nERROR occurrences: {error_count}")
```

Starts/Ends With and String Testing

startswith(prefix) and endswith(suffix) both accept a tuple of alternatives, returning True if the string matches any one of them. This avoids multiple or conditions. The validation methods — isdigit(), isalpha(), isalnum(), isspace(), istitle(), isupper(), islower() — test the content of the string and return a bool.

Example 3 — Real-world application

startswith() with tuple and isdigit/isalnum for API key format validation and file type detection.

```
# startswith with tuple: one call checks multiple prefixes
filenames = ["report_2024.pdf", "data.xlsx", "archive.tar.gz",
             "script.py", "image.PNG", "README.md"]

IMAGE_EXTS  = (".jpg", ".jpeg", ".png", ".gif", ".webp")
OFFICE_EXTS = (".pdf", ".xlsx", ".docx", ".pptx")
```

```
for fname in filenames:
    lower_fname = fname.lower()
    if lower_fname.endswith(IMAGE_EXTS):
        category = "Image"
    elif lower_fname.endswith(OFFICE_EXTS):
        category = "Document"
    elif lower_fname.endswith(".py"):
        category = "Python script"
    else:
        category = "Other"
    print(f"  {fname:<25} -> {category}")

# String content validation
def validate_part_number(pn: str) -> bool:
    """Part numbers: uppercase alphanumeric, 6-12 chars."""
    stripped = pn.strip()
    return (6 <= len(stripped) <= 12 and
            stripped.isalnum() and
            stripped.isupper())

tests = ["ABC123", "PART-001", "abc123", "XY9900", "TOOLONGPARTNO99"]
for t in tests:
    print(f"  {t!r:<20} valid={validate_part_number(t)}")
```

10.3 String Formatting for Tabular Output

f-strings were introduced in Chapter 4 for basic output. Here the focus is on format specifiers that produce precisely aligned tabular output: the kind of formatted tables, reports, and aligned columns that appear in dashboards, invoices, and data exports. Python's format specifier mini-language is the same whether used inside an f-string or with str.format().

Format specifier syntax inside {}: {value:[[fill]align][sign][#][0][width][grouping][.precision][type]}. The most commonly used parts: width (minimum field width), align (< left, > right, ^ centre), fill (character to pad with), , (thousands separator), .N (decimal precision), f/d/s/% (float/integer/string/percentage type).

Tabular Output with Alignment

Example 1 — Introduction

A formatted transaction report with aligned columns using f-string specifiers.

```
transactions = [
    {"id": "TXN-00441", "date": "2024-07-15", "desc": "API Gateway service",
"amount": 1_249.00,  "gst": True},
    {"id": "TXN-00442", "date": "2024-07-15", "desc": "Support plan (annual)",
"amount": 4_800.00,  "gst": True},
    {"id": "TXN-00443", "date": "2024-07-16", "desc": "Domain renewal",
"amount":   89.99,  "gst": False},
    {"id": "TXN-00444", "date": "2024-07-18", "desc": "Additional storage",
"amount":  320.00,  "gst": True},
]

GST_RATE = 0.10
```

```
# Header: left-align text, right-align amounts
  print("ID           Date         Description                     Excl GST     GST
Total")
print("-" * 82)

for t in transactions:
    excl    = t["amount"]
    gst_amt = excl * GST_RATE if t["gst"] else 0
    total   = excl + gst_amt
    print(f"{t['id']:<12} {t['date']:<12} {t['desc']:<26}",
          f"${excl:>9,.2f} ${gst_amt:>7,.2f} ${total:>9,.2f}")

subtotal = sum(t["amount"] for t in transactions)
gst_total = sum(t["amount"] * GST_RATE for t in transactions if t["gst"])
print("-" * 82)
  grand = subtotal + gst_total
  print(f"TOTAL{grand:>77,.2f}")
```

Padding, Fill Characters, and Percentage Formatting

Example 2 — Variation

Fill characters for visual separators, zero-padding for identifiers, and percentage format specifier.

```
# Fill character: pad with a specific character instead of space
  title = "INVOICE TOTAL"
  print(f"{title:=^40}")   # =====INVOICE TOTAL=====
  currency = "AUD"
  print(f"{currency:->10}")             # -------AUD

# Zero-padding: useful for ordered identifiers and codes
for i in range(1, 8):
    order_id = f"ORD-{i:05d}"    # ORD-00001 ... ORD-00007
    print(f"  {order_id}")

# Percentage format: value * 100 then format as %
metrics = [
    ("Cache hit rate",     0.9387),
    ("Error rate",         0.0023),
    ("Uptime",             0.9998),
    ("CPU utilisation",    0.7215),
]
for label, value in metrics:
    bar_len = int(value * 20)
    bar     = "█" * bar_len + "░" * (20 - bar_len)
    print(f"  {label:<22} {bar} {value:>6.2%}")
```

Legacy Formatting: format() and %

str.format() and % formatting are legacy approaches that predate f-strings and appear in older codebases. str.format() uses positional or named placeholders: "Hello, {}!".format(name) or "Hello, {name}!".format(name=name). % formatting uses printf-style codes: "%s is %d years old" % (name, age). Both still work, but f-strings are simpler and faster and should be preferred in new code. Recognising the older forms is important for reading and maintaining existing programs.

```
# str.format() -- positional and named placeholders
msg = "Order {} placed for {} units of {}".format("ORD-441", 12, "Widget A")
print(msg)

# Named placeholders with format()
template = "Dear {name}, your invoice {inv_id} totalling ${amount:.2f} is
due."
print(template.format(name="Chen Wei", inv_id="INV-8820", amount=3_249.95))

# % formatting (legacy, still common in logging and older code)
import logging
level   = "ERROR"
host    = "db-prod-01"
code    = 503
log_msg = "%s from %s: HTTP %d" % (level, host, code)
print(log_msg)
```

10.4 Splitting, Joining, and Partitioning

Breaking strings apart and reassembling them is central to parsing structured text: CSV records, log lines, configuration values, URL paths, and command-line arguments all arrive as strings that must be decomposed before they can be used. Python provides split(), rsplit(), partition(), and the regex-based re.split() for these tasks, and join() for the reverse operation.

split() and rsplit()

split(sep, maxsplit=-1) splits on a separator string and returns a list. With no separator argument, it splits on any whitespace and discards empty strings — the right choice for parsing whitespace-delimited text. rsplit() splits from the right, which is useful when you know the suffix but not the prefix length.

Example 1 — Introduction

Parsing structured log lines using split() with and without a separator.

```
# Log line: "2024-07-15 09:14:02,847 ERROR  auth.service  Failed login: admin"
log_lines = [
    "2024-07-15 09:14:02,847 ERROR  auth.service  Failed login: admin",
    "2024-07-15 09:14:05,112 INFO   api.gateway   Request processed",
    "2024-07-15 09:14:08,334 WARN   cache.layer   Memory threshold: 85%",
]

for line in log_lines:
    # Split on whitespace, maxsplit=4 to keep the message intact
    parts = line.split(None, 4)
    if len(parts) < 5: continue
    date, time_ms, level, module, message = parts
    time_only = time_ms.split(",")[0]
    print(f"  [{level:<5}] {time_only}  {module:<14}  {message}")

# rsplit: extract file extension (may have multiple dots in filename)
filenames = ["report.final.v2.pdf", "data_2024.csv", "archive.tar.gz"]
```

```
for fname in filenames:
    name, _, ext = fname.rsplit(".", 1)
    print(f"  {name!r:<25} ext={ext!r}")
```

partition() and rpartition()

partition(sep) splits the string at the first occurrence of sep and returns a 3-tuple: (before, sep, after). If sep is not found, the full string is in the first element and the other two are empty strings. Unlike split(), partition() always returns exactly three parts and preserves the separator. rpartition() splits at the last occurrence. This is the right tool when you need to split on the first (or last) delimiter without discarding it.

Example 2 — Variation

partition() for HTTP header parsing and URL path splitting.

```
# HTTP header lines: "Content-Type: application/json; charset=utf-8"
headers = [
    "Content-Type: application/json; charset=utf-8",
    "Authorization: Bearer eyJ0eXAiOiJKV1QiLC...",
    "X-Request-ID: 7741-ABCD-9920",
    "Cache-Control: no-cache, no-store",
]

parsed = {}
for header in headers:
    name, sep, value = header.partition(": ")
    if sep:   # sep is empty string if delimiter not found
        parsed[name.strip()] = value.strip()

for k, v in parsed.items():
    print(f"  {k:<20} = {v}")

# rpartition: split a file path at the last separator
path = "/var/log/app/2024/07/service.log"
directory, _, filename = path.rpartition("/")
print(f"\nDir:  {directory}")
print(f"File: {filename}")
```

join() and re.split()

Example 3 — Real-world application

join() to build CSV output and re.split() for multi-delimiter log parsing.

```
import re

# join(): build a CSV row from a list of values
records = [
    ["TXN-001", "2024-07-15", "Widget A", "12", "149.95"],
    ["TXN-002", "2024-07-16", "Gadget, Model B", "3", "89.99"],  # field with
comma
]

for record in records:
```

```
    # Quote fields that contain commas
    quoted = [f'"{f}"' if "," in f else f for f in record]
    csv_line = ",".join(quoted)
    print(csv_line)

# re.split(): split on any of several delimiters
connection_string = "host=db-prod;port=5432 dbname=inventory user=app_user"
tokens = re.split(r"[;\s]+", connection_string)   # semicolons or whitespace
config = {}
for token in tokens:
    if "=" in token:
        key, _, val = token.partition("=")
        config[key] = val
print(f"\nParsed config: {config}")
```

10.5 Searching and Validating Strings

Finding whether a substring exists, where it is, how often it occurs, and whether the entire string matches a required pattern are four distinct operations with four distinct tools. Choosing the right one prevents both errors and unnecessary complexity.

Membership and Position: in, find(), index()

in and not in test membership and return a bool — use them when you only need to know if the substring is present. find(sub, start, end) returns the index of the first occurrence or -1 if absent — use it when you need the position but absence is a normal case. index() behaves like find() but raises ValueError if absent — use it when absence means a programming error and you want the exception. rfind() and rindex() search from the right.

Example 1 — Introduction

in for security header checking, find() for extracting a known field from a structured string.

```
# in: membership check (boolean) -- the simplest and most readable form
response_headers = "Content-Type: application/json\nX-Frame-Options: DENY\nX-
XSS-Protection: 1"

REQUIRED_SECURITY_HEADERS = ["X-Frame-Options", "X-XSS-Protection", "Content-
Security-Policy"]
for header in REQUIRED_SECURITY_HEADERS:
    present = header in response_headers
    status  = "OK" if present else "MISSING"
    print(f"  {header:<30} {status}")

# find(): locate a known token within a string
config_line = "timeout=30 max_retries=5 log_level=WARNING debug=false"

def extract_value(config: str, key: str) -> str | None:
    pos = config.find(key + "=")
    if pos == -1:
        return None
    start = pos + len(key) + 1
    end   = config.find(" ", start)
    return config[start:] if end == -1 else config[start:end]
```

```
for key in ["timeout", "log_level", "missing_key"]:
    print(f"  {key}: {extract_value(config_line, key)!r}")
```

count() and Pattern Validation

count(sub) returns the number of non-overlapping occurrences. For structural validation — checking that a string has exactly one @ sign, or that a version string has exactly two dots — count() is cleaner than split() followed by len().

Example 2 — Variation

count() for structural validation of email addresses and semantic version strings.

```
def validate_email_basic(email: str) -> list[str]:
    """Basic structural email validation without regex."""
    errors = []
    e = email.strip()

    if e.count("@") != 1:
        errors.append("Must contain exactly one @")
        return errors   # Cannot meaningfully continue

    local, domain = e.split("@")
    if not local:
        errors.append("Local part (before @) cannot be empty")
    if "." not in domain:
        errors.append("Domain must contain at least one dot")
    if domain.startswith(".") or domain.endswith("."):
        errors.append("Domain cannot start or end with a dot")
    return errors

def validate_semver(version: str) -> bool:
    """Validate semantic version format: MAJOR.MINOR.PATCH."""
    if version.count(".") != 2:
        return False
    parts = version.split(".")
    return all(p.isdigit() for p in parts)

for email in ["user@example.com", "bad@@example.com", "@nodomain", "no-at-
sign"]:",
    issues = validate_email_basic(email)
    print(f"  {email!r:<30} {'OK' if not issues else issues[0]}")

for v in ["1.2.3", "1.2", "1.2.3.4", "1.2.x"]:",
    print(f"  {v!r:<12} valid={validate_semver(v)}")
```

String Content Testing

The is* methods test the content category of an entire string and return True only if every character satisfies the condition. isdigit() returns True for strings of decimal digit characters. isalpha() for alphabetic. isalnum() for alphanumeric. isspace() for whitespace only. isupper()/islower() for case. These are useful for quick structural checks but have limitations:

isdigit() is True for superscript digits like ^{2}, and none of them accept floats or signed numbers. For numeric conversion, always use try/except with int() or float().

```
# Practical use: classify tokens in a parsed record
tokens = ["ORD-7741", "2024-07-15", "1249", "49.95", "Widget A", "   "]

for token in tokens:
    if token.isspace() or not token.strip():
        category = "empty/whitespace"
    elif token.isdigit():
        category = "integer string"
    elif token.isalpha():
        category = "alphabetic"
    elif token.isalnum():
        category = "alphanumeric"
    else:
        category = "mixed/special"
    print(f"  {token!r:<15} -> {category}")
```

10.6 String Comparison and Sorting

String comparison in Python is case-sensitive and lexicographic (character-by-character by Unicode code point). This produces some surprises: "Z" < "a" is True because uppercase letters have lower code points than lowercase. Normalising case before comparing — with casefold() for maximum robustness — eliminates this class of bug.

Equality and Lexicographic Ordering

Example 1 — Introduction

Case-sensitive vs normalised comparison, and lexicographic sorting of structured identifiers.

```
# Case-sensitive comparison: common source of bugs
status_raw = "Active"
print(status_raw == "active")          # False -- case mismatch
print(status_raw.casefold() == "active") # True  -- normalised

# Sorting strings: lexicographic (dictionary order)
servers = ["db-prod-02", "api-prod-01", "cache-01", "api-staging-01", "db-
prod-01"]

# Default sort: lexicographic, case-sensitive
print(sorted(servers))

# Sort ignoring case
print(sorted(servers, key=str.casefold))

# Sort by service type (prefix before the first hyphen)
print(sorted(servers, key=lambda s: s.partition("-")[0]))
```

Avoiding is for String Comparison

The is operator tests object identity, not value equality. Two string literals with the same content may or may not refer to the same object in memory, depending on Python's string interning optimisation. is may return True for short literal strings (because Python interns them) but False for dynamically created strings with the same value. Always use == to compare string content. Use is only to test against None or other singletons.

Example 2 — Variation

Demonstrating why is is unreliable for string comparison and should never be used for it.

```
# is works for interned literals (accident of implementation)
a = "active"
b = "active"
print(a is b)     # True -- CPython interns short literals (do not rely on
this)

# is breaks for dynamically created strings
status = "act" + "ive"   # created at runtime
print(status is "active")  # False (or True -- implementation-defined)
print(status == "active")  # Always True -- correct comparison

# The only correct check for None:
value = None
if value is None:     # correct
    print("No value")
if value == None:     # works but is not idiomatic; use is
    print("No value")
```

Normalisation and Deduplication

Example 3 — Real-world application

Normalising and deduplicating a list of company names from different data sources.

```
raw_names = [
    "   Acme Corporation   ",
    "ACME CORPORATION",
    "acme corporation",
    "Acme Corp.",
    "Global Widgets Pty Ltd",
    "GLOBAL WIDGETS PTY LTD",
    "global widgets",
]

def normalise_company(name: str) -> str:
    """Produce a normalised key for deduplication."""
    # Strip, casefold, remove common suffixes and punctuation
    n = name.strip().casefold()
    for suffix in [" pty ltd", " pty. ltd.", " ltd", " corp.", " corporation",
" inc."]:",
        if n.endswith(suffix):
            n = n[:-len(suffix)].strip()
            break
    return n.replace(".", "").strip()

seen     = {}
for name in raw_names:
    key = normalise_company(name)
    if key not in seen:
        seen[key] = name.strip()    # keep first occurrence
```

```
print("Deduplicated names:")
for key, canonical in seen.items():
    print(f"  {canonical!r:<35} (key: {key!r})")
```

10.7 Common String Processing Patterns

The string methods and techniques covered in this chapter compose into higher-level patterns that appear repeatedly in real programs: log parsing, slug generation for URLs, text normalisation, and building formatted output from templates. This section demonstrates each as a compact, self-contained function.

Log Line Parsing

Example 1 — Introduction

A reusable log parser that extracts structured fields from a fixed-format log line.

```
from datetime import datetime

def parse_log_line(line: str) -> dict | None:
    """
    Parse lines of the form:
    "2024-07-15T09:14:02 ERROR  module.name  Message text here"
    Returns a dict or None if the line does not match.
    """
    line = line.strip()
    if not line:
        return None

    parts = line.split(None, 3)   # split on whitespace, keep message intact
    if len(parts) < 3:
        return None

    try:
        timestamp = datetime.fromisoformat(parts[0])
    except ValueError:
        return None

    level  = parts[1].upper()
    module = parts[2] if len(parts) > 3 else ""
    message = parts[3] if len(parts) > 3 else parts[2]

    return {"timestamp": timestamp, "level": level,
            "module": module, "message": message}

raw_lines = [
    "2024-07-15T09:14:02 ERROR  auth.service  Failed login attempt",
    "2024-07-15T09:14:05 INFO   api.gateway   Request processed OK",
    "",
    "not a valid log line",
]

errors = [r for line in raw_lines if (r := parse_log_line(line)) and
r["level"] == "ERROR"]
```

```
print(f"Errors found: {len(errors)}")
for e in errors:
    print(f"  [{e['timestamp'].strftime('%H:%M:%S')}] {e['module']}: {e['message']}")
```

Slug Generation

Example 2 — Variation

Converting arbitrary text to a URL-safe slug: a chain of string operations.

```
import re

def slugify(text: str, max_length: int = 80) -> str:
    """
    Convert text to a URL-safe slug.
    "Q3 2024: Revenue Report (Final)" -> "q3-2024-revenue-report-final"
    """
    # Casefold and strip
    slug = text.strip().casefold()
    # Replace common punctuation and spaces with hyphens
    slug = re.sub(r"[^a-z0-9]+", "-", slug)
    # Remove leading/trailing hyphens
    slug = slug.strip("-")
    # Truncate at a word boundary to avoid cutting mid-word
    if len(slug) > max_length:
        slug = slug[:max_length].rsplit("-", 1)[0]
    return slug

titles = [
    "Q3 2024: Revenue Report (Final)",
    "API Gateway -- Configuration Guide v2.1",
    "FAQ: What's new in Python 3.12?",
    "A" * 100,    # very long input
]
for t in titles:
    print(f"  {t!r[:40]:<42} -> {slugify(t)!r}")
```

Text Normalisation Pipeline

Example 3 — Real-world application

A normalisation pipeline for cleaning free-text customer feedback before analysis.

```
import re

def normalise_feedback(text: str) -> str:
    """
    Normalise free-text feedback for analysis:
    - Strip leading/trailing whitespace
    - Collapse internal whitespace runs to single spaces
    - Remove non-printable/control characters
    - Normalise common contractions
    - Truncate to a maximum length
    """
    MAX_LEN = 500
```

```
    CONTRACTIONS = {"dont": "do not", "cant": "cannot",
                    "wont": "will not", "didnt": "did not"}

    text = text.strip()
    text = re.sub(r"[^\x20-\x7E]+", " ", text)   # remove non-printable
    text = re.sub(r"\s+", " ", text)               # collapse whitespace

    # Expand contractions
    words = text.lower().split()
    words = [CONTRACTIONS.get(w.replace("'", ""), w) for w in words]
    text  = " ".join(words)

    return text[:MAX_LEN] if len(text) > MAX_LEN else text

samples = [
    "  The service  was GREAT!!  ",
    "I cant believe how fast it was\x00\x01",
    "Dont use this. Wont recommend.",
]
for s in samples:
    print(f"  Input:  {s!r}")
    print(f"  Output: {normalise_feedback(s)!r}\n")
```

Chapter 10 Summary

- Strings are immutable ordered sequences of Unicode characters. Indexing (s[i]) accesses single characters; slicing (s[start:stop:step]) extracts substrings; negative indices count from the end. String operations always produce new strings. Raw strings (r"...") suppress backslash interpretation. Triple-quoted strings span multiple lines.
- Key case methods: lower() for simple normalisation, casefold() for robust case-insensitive comparison including non-ASCII characters, upper() for display, title() for title-casing words. strip() / lstrip() / rstrip() remove whitespace or a specified character set. replace(old, new, count) substitutes substrings. Method calls chain: s.strip().lower().replace("-", "_").
- f-string format specifiers: {value:[[fill]align][width][,][.precision][type]}. Common: :<width (left-align), :>width (right-align), :^width (centre), :, (thousands separator), :.2f (2 decimal places), :.2% (percentage), :05d (zero-pad integer). Fill character replaces space: {:=^40} centres with = fill. str.format() and % are legacy alternatives still present in older codebases.
- split(sep, maxsplit) breaks a string into a list; with no separator it splits on any whitespace and discards empty strings. rsplit() splits from the right. partition(sep) returns (before, sep, after) as a 3-tuple, preserving the separator; rpartition() from the right. join(iterable) is the reverse of split(). re.split(pattern) splits on a regex pattern for multi-delimiter text.
- Use in / not in for membership (returns bool). find(sub) returns the index of the first occurrence or -1. index(sub) returns the index or raises ValueError. count(sub) counts non-overlapping occurrences. startswith() and endswith() accept tuples of alternatives. rfind() and rindex() search from the right.
- String content testing: isdigit() (decimal digit characters), isalpha() (alphabetic), isalnum() (alphanumeric), isspace() (whitespace only), isupper() / islower(). These return True only if every character satisfies the condition. For numeric validation, always use try/except with int() or float() — isdigit() fails for negative numbers, floats, and scientific notation.

- String comparison is case-sensitive and lexicographic. Always use == to compare string content. Never use is for equality — it tests object identity, not value, and produces unpredictable results for dynamically created strings. Use casefold() before comparing or sorting strings that may differ only in case. Common patterns: log parsing with split(), slug generation with re.sub(), deduplication with casefold() normalisation.

Guided Exercises

Complete the following exercises in VS Code, IDLE, or your preferred Python environment. Each exercise targets a specific concept from this chapter.

1. Write a log file parser. Given a list of log lines in the format "TIMESTAMP LEVEL MODULE MESSAGE", use split() with maxsplit to extract each field. Validate that the timestamp matches the expected format using datetime.fromisoformat(). Collect all ERROR and CRITICAL records and print a formatted summary report showing count by level, the first and last error timestamp, and the unique modules that reported errors.

2. Build a CSV field cleaner. Given a list of raw string fields (with leading/trailing whitespace, inconsistent casing, currency symbols, and some empty values), apply a cleaning pipeline: strip whitespace, normalise case for categorical fields using casefold(), convert price strings (which may contain "$", commas, or be empty) to floats using try/except, and validate that required fields are non-empty. Print a before/after comparison for each field.

3. Write a formatted invoice renderer. Given a list of line items (description, quantity, unit price), produce a formatted invoice using f-string specifiers: descriptions left-aligned in a 30-char field, quantities right-aligned in 6 chars, unit prices and totals right-aligned with two decimal places and thousands separators. Include a separator line, subtotal, GST (10%), and grand total. All monetary values must align vertically.

4. Implement a string validation library with four functions: validate_email(s) using count() and partition(); validate_semver(s) using count() and split(); validate_identifier(s) using isalnum() and a custom allowed-character check; validate_phone_au(s) to check Australian phone number formats. Each function should return a tuple of (is_valid: bool, error: str | None). Test each with at least six inputs including edge cases.

5. Write a company name normaliser and deduplicator. Given a list of company names from different sources (with varying capitalisation, spacing, and common suffixes like "Pty Ltd", "Corp.", "Inc."), produce a normalised key for each using casefold() and replace(). Group names by their normalised key and print the groups, showing which names were considered duplicates. Include at least 15 input names with at least 4 groups of duplicates.

Challenge Tasks

The following tasks are open-ended. No hints are provided. They require you to design a solution, not just apply a pattern.

1. Build a configuration file parser that handles a simple key=value format with comments (lines starting with #), sections ([section_name]), type inference (integers, floats, booleans, and strings), and multi-line values (continued with a trailing backslash). The parser should return a nested dictionary keyed by section, then by key. Implement a get(section, key, default=None) method with type-appropriate defaults. Test with a realistic configuration file covering all features, and write a validator that checks that required keys are present and their values are within acceptable ranges.

2. Implement a text templating engine. Templates use {variable_name} for substitution, {variable_name!upper} / {variable_name!lower} / {variable_name!title} for transformed substitution, and {variable_name:>20} for width-formatted substitution. The engine should raise a descriptive error for unknown variables, handle missing variables with a configurable default, and support escape of literal braces with doubled braces {{ and }}. Demonstrate with at least three different templates including one that uses all features.

3. Write a log analytics tool. Given a log file (or a simulated list of at least 50 log lines with realistic variety), the tool should: parse each line into structured fields; compute summary statistics (count by level, count by module, errors per minute, the 10 most common error messages); identify any recurring patterns in error messages using substring matching; format the entire analysis as a readable report with aligned columns and section headers. Make the tool robust to malformed lines by counting and reporting them separately.

CHAPTER 11 Creating Lists

Think, Code, Create – A Python Programming Journey

The list is Python's most versatile built-in data structure. It is an ordered, mutable sequence that can hold any mix of types, grow and shrink dynamically, and be iterated, sliced, sorted, searched, and copied with a rich set of built-in tools. Most programs that work with collections of data — transaction records, log entries, configuration parameters, sensor readings, API results — reach for a list as the first container.

This chapter covers lists comprehensively: creating and indexing them, the full suite of mutating methods, sorting and searching, nested lists for multi-dimensional data, common structural patterns including lists as stacks and queues, in-place modification with slicing, and the subtle but important distinction between shallow and deep copies. All examples use realistic data from practical domains.

11.1 Introduction to Lists

A list is created with square brackets containing comma-separated values. The values may be of any type — integers, floats, strings, booleans, other lists, or any Python object — and different types can coexist in the same list. Lists are zero-indexed: the first element is at index 0, the last at index -1. Lists are mutable: elements can be replaced, added, or removed after creation.

Creating and Indexing

Example 1 — Introduction

Creating a list of server records and accessing elements by positive and negative index.

```
# A list of server status records
servers = [
    {"host": "api-prod-01",  "cpu": 62, "status": "healthy"},
    {"host": "db-primary",   "cpu": 88, "status": "degraded"},
    {"host": "cache-01",     "cpu": 34, "status": "healthy"},
    {"host": "worker-01",    "cpu": 97, "status": "critical"},
]

# Positive indexing (from the front)
print(servers[0]["host"])   # "api-prod-01" -- first element
print(servers[2]["cpu"])    # 34            -- third element

# Negative indexing (from the back)
print(servers[-1]["status"])  # "critical"  -- last element
print(servers[-2]["host"])    # "cache-01"  -- second-to-last

# Length
print(f"{len(servers)} servers monitored")
```

```
# Membership test
hosts = [s["host"] for s in servers]
print("api-prod-01" in hosts)   # True
print("db-replica" in hosts)    # False
```

Mutability

Unlike strings (which are immutable), lists can be modified in place. Assigning to an index replaces the element at that position. This mutability is a core feature — it is what makes lists suitable for representing state that changes over time. However, it also means that multiple variables pointing to the same list will all see any modifications; section 11.7 covers this carefully.

Example 2 — Variation

In-place element replacement and the difference between list mutation and reassignment.

```
alerts = ["INFO: startup complete", "WARN: high memory", "ERROR: disk full"]

# Replace an element by index
alerts[1] = "WARN: memory at 94%"   # updated warning
print(alerts)

# Mutation is visible through all references to the same list
active_alerts = alerts      # NOT a copy -- same list object
alerts[2] = "CRITICAL: disk quota exceeded"
print(active_alerts[2])     # "CRITICAL: disk quota exceeded" -- both changed

# To work independently, make a copy (covered in section 11.7)
snapshot = alerts.copy()    # shallow copy
alerts[0] = "INFO: restart initiated"
print(snapshot[0])          # "INFO: startup complete" -- snapshot unaffected
```

Slicing

List slicing extracts a new list using the syntax list[start:stop:step]. The start index is inclusive, stop is exclusive. Omitting start defaults to 0; omitting stop defaults to len(list). A negative step reverses the direction. Slicing never raises an IndexError — out-of-range indices are clamped silently.

Example 3 — Real-world application

Slicing for pagination, reversing, and sampling from a transaction log.

```
transactions = [
    {"id": f"T-{i:04d}", "amount": (i * 17 % 500) + 10}
    for i in range(1, 21)    # 20 simulated transactions
]

PAGE_SIZE = 5

# Pagination: page 2 (0-indexed)
page = 2
start = page * PAGE_SIZE
page_records = transactions[start:start + PAGE_SIZE]
```

```
print(f"Page {page}: {[r['id'] for r in page_records]}")

# Most recent 3 transactions (last 3)
recent = transactions[-3:]
print(f"Recent: {[r['id'] for r in recent]}")

# Every other transaction (step 2)
sample = transactions[::2]
print(f"Sample ({len(sample)} records): {[r['id'] for r in sample]}")

# Reversed order
reversed_txns = transactions[::-1]
print(f"First in reversed: {reversed_txns[0]['id']}")
```

11.2 List Methods

Python lists have eleven built-in methods that cover all the common mutation operations: adding elements, removing elements, finding positions, counting, sorting, reversing, and copying. All mutation methods modify the list in place and return None (not a new list) — a common source of bugs when developers write new_list = my_list.sort() expecting a sorted list back. Use sorted() (section 11.3) to obtain a new sorted list without modifying the original.

Adding Elements: append(), insert(), extend()

append(x) adds a single element to the end. insert(i, x) inserts x before the element at index i; index 0 inserts at the beginning, len(list) is equivalent to append. extend(iterable) adds every element from the iterable to the end — it is equivalent to list += iterable. The critical distinction: append([1,2,3]) adds one element (a list); extend([1,2,3]) adds three elements.

Example 1 — Introduction

append(), insert(), and extend() building a queue of processing jobs.

```
job_queue = ["JOB-001", "JOB-002", "JOB-003"]

# append: add a job to the end of the queue
job_queue.append("JOB-004")
print(f"After append: {job_queue}")

# insert: add an urgent job at the front (index 0)
job_queue.insert(0, "JOB-URGENT")
print(f"After insert: {job_queue}")

# extend: add multiple jobs from another list
backlog = ["JOB-005", "JOB-006", "JOB-007"]
job_queue.extend(backlog)
print(f"After extend: {len(job_queue)} jobs in queue")

# Common mistake: append vs extend
queue_a = ["JOB-001"]
queue_a.append(["JOB-002", "JOB-003"])  # adds ONE element: a list
print(f"append list: {queue_a}")         # ["JOB-001", ["JOB-002", "JOB-003"]]

queue_b = ["JOB-001"]
```

```
queue_b.extend(["JOB-002", "JOB-003"]) # adds TWO elements
print(f"extend list: {queue_b}")         # ["JOB-001", "JOB-002", "JOB-003"]
```

Removing Elements: remove(), pop(), clear()

remove(x) removes the first occurrence of x; raises ValueError if not found. pop(i) removes and returns the element at index i (default: last element). clear() empties the list. For safe removal when the element may not be present, check with in before calling remove(), or use a try/except.

Example 2 — Variation

Removing completed and cancelled jobs from a processing queue.

```
jobs = [
    {"id": "JOB-001", "status": "complete"},
    {"id": "JOB-002", "status": "running"},
    {"id": "JOB-003", "status": "cancelled"},
    {"id": "JOB-004", "status": "running"},
    {"id": "JOB-005", "status": "complete"},
]

# pop(): remove and return by index (stack behaviour -- from end)
last_job = jobs.pop()
print(f"Popped: {last_job['id']} ({last_job['status']})")

# pop(0): remove from front (queue behaviour)
first_job = jobs.pop(0)
print(f"Dequeued: {first_job['id']} ({first_job['status']})")

# Remove all completed/cancelled jobs (iterate over a copy)
to_remove = [j for j in jobs if j["status"] in ("complete", "cancelled")]
for job in to_remove:
    jobs.remove(job)

print(f"Running jobs: {[j['id'] for j in jobs]}")
```

Other Methods: reverse(), copy(), index(), count()

Example 3 — Real-world application

reverse(), copy(), index(), and count() in an event log context.

```
events = ["startup", "connect", "login", "error", "retry", "login", "logout"]

# reverse(): in-place reversal (returns None)
events.reverse()
print(f"Reversed: {events[:3]}...")   # most recent first

# copy(): shallow copy (safe to modify independently)
snapshot = events.copy()
events.append("reconnect")
print(f"Snapshot unchanged: {len(snapshot)} events, live: {len(events)}")

# index(x): find first occurrence (raises ValueError if absent)
events_fwd = ["startup", "connect", "login", "error", "login"]
```

```
try:
    pos = events_fwd.index("login")
    print(f"First 'login' at position {pos}")
except ValueError:
    print("Event not found")

# count(x): how many times x appears
login_count = events_fwd.count("login")
error_count = events_fwd.count("error")
print(f"Login events: {login_count}, Error events: {error_count}")
```

11.3 Searching and Sorting

Searching a list for a value or position, and sorting a list by a key, are two of the most frequent operations in data processing. Python provides both in-place and non-destructive variants: list.sort() modifies the list directly and returns None; sorted() returns a new sorted list and leaves the original unchanged. This distinction matters when the original order must be preserved.

sort() and sorted() with key=

Both sort() and sorted() accept a key= argument: a function applied to each element before comparison. The element itself is not changed — only its sort key. Common key functions include str.lower for case-insensitive string sorting, len for length, and lambda expressions for attribute access on dictionaries or objects. The reverse=True argument reverses the sort order.

Example 1 — Introduction

Sorting a list of server records by different keys: CPU usage, hostname, and status.

```
servers = [
    {"host": "api-prod-01",  "cpu": 62, "status": "healthy"},
    {"host": "db-primary",   "cpu": 88, "status": "degraded"},
    {"host": "cache-01",     "cpu": 34, "status": "healthy"},
    {"host": "worker-01",    "cpu": 97, "status": "critical"},
]

# sorted(): returns new list, original unchanged
by_cpu = sorted(servers, key=lambda s: s["cpu"], reverse=True)
print("By CPU (highest first):")
for s in by_cpu:
    print(f"  {s['host']:<15} {s['cpu']}%")

# sort(): in-place, returns None
servers.sort(key=lambda s: (s["status"], s["host"]))
print("\nBy status then host:")
for s in servers:
    print(f"  {s['status']:<10} {s['host']}")

# Case-insensitive sort of strings
hostnames = ["Worker-01", "api-prod-01", "Cache-01", "DB-Primary"]
print("\nCase-insensitive:", sorted(hostnames, key=str.casefold))
```

Stable Sorting and Multi-Key Sorting

Python's sort is stable: elements that compare equal retain their original relative order. This makes it easy to sort by multiple criteria — sort by the secondary key first, then by the primary key. The result respects both. Alternatively, use a tuple as the key to sort by multiple fields simultaneously.

Example 2 — Variation

Multi-key sorting of transactions: by status (primary) then by amount descending (secondary).

```
transactions = [
    {"id": "T-003", "status": "pending",   "amount": 320.00},
    {"id": "T-001", "status": "completed", "amount": 1_249.00},
    {"id": "T-005", "status": "completed", "amount":   89.99},
    {"id": "T-002", "status": "pending",   "amount":  450.00},
    {"id": "T-004", "status": "failed",    "amount":  175.00},
]

# Sort by status, then by amount descending within each status
# Tuples compare element-by-element; negate amount to reverse just that key
sorted_txns = sorted(
    transactions,
    key=lambda t: (t["status"], -t["amount"])
)

for t in sorted_txns:
    print(f"  [{t['status']:<10}] {t['id']}  ${t['amount']:>8,.2f}")
```

Searching: in, index(), and Linear vs Binary Search

For membership testing, in runs in O(n) time (linear scan). For sorted lists where performance matters with large datasets, the bisect module provides O(log n) binary search. Use in for small lists and one-off checks; use bisect for repeated lookups in large sorted lists.

Example 3 — Real-world application

Comparing in with bisect for efficient sorted-list lookups in a price tier system.

```
import bisect

# Price tiers: each threshold marks the start of the next tier
TIER_THRESHOLDS = [0, 100, 500, 2_000, 10_000]  # must be sorted
TIER_NAMES      = ["Micro", "Basic", "Standard", "Business", "Enterprise"]

def get_tier(amount: float) -> str:
    """Return the tier name for a given transaction amount."""
    # bisect_right finds the insertion point, which maps to the tier
    idx = bisect.bisect_right(TIER_THRESHOLDS, amount) - 1
    return TIER_NAMES[min(idx, len(TIER_NAMES) - 1)]

test_amounts = [50, 99, 100, 450, 500, 1_999, 2_000, 15_000]
for amount in test_amounts:
    print(f"  ${amount:>7,}  ->  {get_tier(amount)}")
```

11.4 Nested Lists and Multi-Level Data

A list whose elements are themselves lists is called a nested list. Nested lists model two-dimensional data — tables, matrices, grids, and any structure where rows contain columns. Deeper nesting models higher-dimensional data but is increasingly hard to read; dictionaries or dedicated data classes are usually preferable beyond two levels.

Creating and Accessing Nested Lists

Example 1 — Introduction

A weekly shift schedule as a nested list: outer list is days, inner is shifts.

```
# Shift schedule: roster[day][shift] -> employee name
DAYS   = ["Mon", "Tue", "Wed", "Thu", "Fri"]
SHIFTS = ["Morning", "Afternoon", "Evening"]

roster = [
    ["Chen Wei",     "Amara Okafor", "Marcus Silva"],    # Monday
    ["Linh Nguyen", "Chen Wei",      "Priya Sharma"],    # Tuesday
    ["Amara Okafor","Linh Nguyen",  "Chen Wei"],         # Wednesday
    ["Marcus Silva","Priya Sharma", "Amara Okafor"],     # Thursday
    ["Priya Sharma","Marcus Silva", "Linh Nguyen"],      # Friday
]

# Access: roster[day_index][shift_index]
print(f"Wednesday Evening: {roster[2][2]}")
print(f"Friday Morning:    {roster[4][0]}")

# Print the full schedule
  header = "Day   "
  for shift in SHIFTS: header += f"  {shift:<16}"
  print("\n" + header)
for shift in SHIFTS: print(f"  {shift:<16}", end="")
print()
print("-" * 60)
for day, shifts in zip(DAYS, roster):
    print(f"{day:<6}", end="")
    for employee in shifts:
        print(f"  {employee:<16}", end="")
    print()
```

Modifying Nested Lists

Nested lists are mutable at every level. You can replace an element within a sub-list, append to a sub-list, or replace an entire sub-list. When creating a 2D list programmatically, use a list comprehension rather than list multiplication — the multiplication shortcut creates aliased references, not independent inner lists.

Example 2 — Variation

The aliasing trap in nested list creation, and the correct list comprehension approach.

```
ROWS, COLS = 3, 4

# WRONG: all rows are the same object
bad_grid = [[0] * COLS] * ROWS
```

```
bad_grid[0][0] = 99
print("Bad grid:", bad_grid)   # ALL rows changed -- aliased!

# CORRECT: list comprehension creates independent inner lists
grid = [[0] * COLS for _ in range(ROWS)]
grid[0][0] = 99
print("Good grid:", grid)      # only row 0, col 0 changed

# Updating a specific cell in the roster
roster = [
    ["Chen Wei",    "Amara Okafor", "Marcus Silva"],
    ["Linh Nguyen", "Chen Wei",     "Priya Sharma"],
]
roster[0][1] = "Kai Johnson"   # Wednesday afternoon cover
print(f"Updated Wednesday afternoon: {roster[0][1]}")
```

Flattening and Comprehensions on Nested Lists

Example 3 — Real-world application

Flattening a nested structure and filtering across all sub-lists with comprehensions.

```
# Monthly sales by region: sales[month][region] = revenue
monthly_by_region = [
    [142_000, 98_000, 71_000],   # Jan: [NSW, VIC, QLD]
    [138_500, 102_400, 68_200],  # Feb
    [155_000, 110_200, 82_000],  # Mar
]
REGIONS = ["NSW", "VIC", "QLD"]

# Flatten to one list of all monthly figures
all_figures = [rev for month in monthly_by_region for rev in month]
print(f"Total figures: {len(all_figures)}")
print(f"Grand total:   ${sum(all_figures):,.0f}")
print(f"Best month-region: ${max(all_figures):,.0f}")

# Column totals: sum each region across all months
for i, region in enumerate(REGIONS):
    total = sum(monthly_by_region[m][i] for m in
range(len(monthly_by_region)))
    print(f"  {region}: ${total:,.0f}")
```

11.5 List Operations and Structural Patterns

Beyond the individual methods, lists support several operators and participate in common structural patterns that appear repeatedly in real programs: using lists as stacks, as queues (with the collections.deque alternative), and as accumulators for building results incrementally.

Operators: Concatenation, Repetition, Membership

Example 1 — Introduction

*The +, *, in, and not in operators on lists in a data pipeline context.*

```
approved_regions = ["AU", "NZ", "SG"]
```

```python
pending_regions  = ["JP", "KR"]

# + concatenation: creates a new list
all_regions = approved_regions + pending_regions
print(f"All regions: {all_regions}")

# * repetition: creates a new list with repeated elements
# Useful for initialising a fixed-size list
error_counts = [0] * len(all_regions)  # [0, 0, 0, 0, 0]
print(f"Error counts init: {error_counts}")

# in / not in: membership test (O(n) linear scan)
request_region = "AU"
if request_region in approved_regions:
    print(f"Region {request_region} approved")

# += is equivalent to extend() -- modifies in place
approved_regions += ["HK", "TW"]
print(f"After +=: {approved_regions}")
```

List as a Stack (LIFO)

A stack is a last-in, first-out (LIFO) structure. append() pushes to the top; pop() removes from the top. Both are O(1) operations on a list. This pattern is used for undo systems, function call tracking, expression parsing, and depth-first traversal.

Example 2 — Variation

A deployment rollback system implemented as a stack of configuration snapshots.

```python
class DeploymentStack:
    """Track configuration snapshots for rollback."""
    def __init__(self):
        self._stack = []

    def push(self, config: dict):
        self._stack.append(config)
        print(f"  Snapshot saved: v{len(self._stack)}")

    def rollback(self) -> dict | None:
        if not self._stack:
            print("  Nothing to roll back to.")
            return None
        config = self._stack.pop()
        print(f"  Rolled back. Snapshots remaining: {len(self._stack)}")
        return config

    def peek(self) -> dict | None:
        return self._stack[-1] if self._stack else None

stack = DeploymentStack()
stack.push({"version": "1.0.0", "replicas": 3})
stack.push({"version": "1.1.0", "replicas": 5})
stack.push({"version": "1.2.0-beta", "replicas": 5})
print(f"Current: {stack.peek()}")
stack.rollback()
```

```
print(f"After rollback: {stack.peek()}")
```

List as a Queue and collections.deque

A queue is a first-in, first-out (FIFO) structure. Appending to the end and popping from the front with pop(0) works on a list, but pop(0) is O(n) because every element must shift left. For a high-volume queue, use collections.deque, which supports O(1) append to either end and O(1) pop from either end.

Example 3 — Real-world application

A task queue implemented with deque for O(1) enqueue and dequeue operations.

```
from collections import deque

task_queue = deque()

# Enqueue tasks (append to right)
for task_id in ["TASK-001", "TASK-002", "TASK-003", "TASK-004"]:",
    task_queue.append({"id": task_id, "status": "pending"})
    print(f"  Queued: {task_id}")

print(f"Queue depth: {len(task_queue)}")

# Dequeue tasks (pop from left -- FIFO)
while task_queue:
    task = task_queue.popleft()
    print(f"  Processing: {task['id']}")

# deque also supports maxlen for a fixed-size sliding window
recent_errors = deque(maxlen=5)
for i in range(8):
    recent_errors.append(f"ERR-{i}")
print(f"Last 5 errors: {list(recent_errors)}")
```

11.6 List Slicing for In-Place Modification

Slice assignment allows replacing, inserting, or deleting a range of elements in a list without a loop. The left side is a slice; the right side is any iterable. The two sides do not need to be the same length — the list adjusts its size automatically. This is more concise and often faster than the equivalent loop-with-pop-and-insert approach.

Replace, Insert, and Delete by Slice

Example 1 — Introduction

Slice assignment for bulk replacement, insertion, and deletion in a configuration list.

```
config_lines = [
    "host = db-prod-01",
    "port = 5432",
    "# legacy timeout settings",
    "timeout = 30",
    "retry = 3",
```

```
    "pool_size = 10",
]

# Replace a slice: swap lines 2-4 with new values
config_lines[2:5] = [
    "# updated timeout settings",
    "connect_timeout = 10",
    "query_timeout   = 60",
    "idle_timeout    = 300",
]
print("After replace:", len(config_lines), "lines")

# Insert at position 2 (without replacing): assign empty slice
config_lines[2:2] = ["# connection settings"]

# Delete lines 4-5: assign empty list to slice
config_lines[4:6] = []

for line in config_lines:
    print(f"  {line}")
```

Slice Deletion and Reversal

Example 2 — Variation

Using del with a slice for deletion and slice assignment for in-place batch update.

```
readings = [18.2, 19.1, -999, 20.4, -999, 21.0, 19.8, -999, 22.1]
SENTINEL  = -999

# Remove all sentinel values in place using slice assignment
  # (cannot modify a list while iterating it -- build indices first)
sentinel_indices = [i for i, v in enumerate(readings) if v == SENTINEL]

# Remove in reverse order to avoid index shifting
for i in reversed(sentinel_indices):
    del readings[i]

print(f"Clean readings: {readings}")

# Bulk replace via slice: apply calibration offset to a range
CALIBRATION_OFFSET = 0.3
readings[2:5] = [round(v + CALIBRATION_OFFSET, 1) for v in readings[2:5]]
print(f"After calibration: {readings}")
```

Practical Slicing Patterns

Example 3 — Real-world application

Common slicing idioms: take first/last N, drop first/last N, every Nth, and rotate.

```
scores = [88, 92, 71, 95, 67, 84, 78, 90, 73, 85]

top_3       = sorted(scores, reverse=True)[:3]
bottom_3    = sorted(scores)[:3]
```

```python
without_extremes = sorted(scores)[1:-1]   # drop lowest and highest
every_third   = scores[::3]

# Rotate left by N positions
def rotate_left(lst, n):
    n = n % len(lst)
    return lst[n:] + lst[:n]

priority_queue = ["critical", "high", "medium", "low", "low", "medium"]
rotated = rotate_left(priority_queue, 2)

print(f"Top 3:            {top_3}")
print(f"Bottom 3:         {bottom_3}")
print(f"Without extremes: {without_extremes}")
print(f"Every third:      {every_third}")
print(f"Rotated left 2:   {rotated}")
```

11.7 Copying Lists: Shallow vs Deep

Because lists are mutable and assignment only copies a reference (not the object), working with copies of lists requires explicit thought. Python provides three shallow copy methods and one deep copy mechanism, and choosing the wrong one is a common source of bugs where modifying one list unexpectedly changes another.

The Reference Problem

Example 1 — Introduction

Demonstrating that assignment copies the reference, not the list.

```python
config_a = ["host=db-prod-01", "port=5432", "pool=10"]

# Assignment: both names point to the SAME list object
config_b = config_a
config_b[0] = "host=db-staging-01"

print(f"config_a: {config_a}")  # also changed!
print(f"Same object: {config_a is config_b}")

# Any of these three create a SHALLOW COPY (independent outer list)
config_c = config_a.copy()     # method
config_d = config_a[:]         # slice
config_e = list(config_a)      # constructor

config_c[0] = "host=localhost"
print(f"config_a unchanged: {config_a[0]}")
print(f"config_c changed:   {config_c[0]}")
```

Shallow Copy Limitations with Nested Lists

A shallow copy creates a new outer list, but the inner objects (sub-lists, dicts, or any mutable objects) are shared between the original and the copy. Modifying a nested mutable object through either list affects both. Use copy.deepcopy() when the nested contents must also be independent.

Example 2 — Variation

Shallow copy sharing inner lists, and deepcopy() providing full independence.

```
import copy

team_schedules = [
    ["Mon 09:00", "Mon 14:00"],    # Team Alpha
    ["Tue 10:00", "Tue 15:00"],    # Team Beta
]

# Shallow copy: inner lists are shared
shallow = team_schedules.copy()
shallow[0].append("Mon 19:00")   # modifies team_schedules[0] too!

print("After shallow copy modification:")
print(f"  Original Team Alpha: {team_schedules[0]}")
print(f"  Shallow Team Alpha:  {shallow[0]}")

# Deep copy: completely independent
team_schedules2 = [["Mon 09:00", "Mon 14:00"], ["Tue 10:00", "Tue 15:00"]]
deep = copy.deepcopy(team_schedules2)
deep[0].append("Mon 19:00")

print("After deep copy modification:")
print(f"  Original Team Alpha: {team_schedules2[0]}")
print(f"  Deep Team Alpha:     {deep[0]}")
```

When to Use Each

Use direct assignment when you intentionally want two names to refer to the same list. Use a shallow copy (copy(), [:], or list()) when the list contains only immutable elements (strings, numbers, tuples) and you need an independent outer list. Use deepcopy() when the list contains mutable nested objects that must be fully independent. In practice, most shallow copies of lists of primitives or strings are safe — the nested-mutable scenario is the one requiring care.

Example 3 — Real-world application

Choosing the right copy strategy when snapshotting a list of mutable configuration objects.

```
import copy

configs = [
    {"service": "api",   "replicas": 3, "env_vars": {"LOG_LEVEL": "INFO"}},
    {"service": "cache", "replicas": 2, "env_vars": {"MAX_MEM": "512m"}},
]

# Shallow copy: dict objects are still shared
staging_configs = configs.copy()
staging_configs[0]["replicas"] = 1   # also changes configs[0]!
staging_configs[0]["env_vars"]["LOG_LEVEL"] = "DEBUG"  # also changes original

# Deep copy: fully independent -- correct for snapshot
configs2 = [
```

```
    {"service": "api",   "replicas": 3, "env_vars": {"LOG_LEVEL": "INFO"}},
    {"service": "cache", "replicas": 2, "env_vars": {"MAX_MEM": "512m"}},
]
snapshot = copy.deepcopy(configs2)
snapshot[0]["replicas"] = 1
snapshot[0]["env_vars"]["LOG_LEVEL"] = "DEBUG"

print(f"Original replicas: {configs2[0]['replicas']}  (unchanged)")
print(f"Snapshot replicas: {snapshot[0]['replicas']}  (modified
independently)")
```

Chapter 11 Summary

- Lists are ordered, mutable sequences created with []. Elements can be of any type and are accessed by zero-based positive indices or negative indices from the end. Slicing (list[start:stop:step]) extracts a new list without modifying the original; out-of-range indices are clamped silently.
- Mutation methods: append(x) adds to the end; insert(i, x) inserts before index i; extend(iterable) adds all elements from the iterable. remove(x) removes the first occurrence (ValueError if absent); pop(i) removes and returns by index (default: last); clear() empties the list. All mutation methods return None, not a new list.
- index(x) returns the position of the first occurrence (ValueError if absent); count(x) counts occurrences; reverse() reverses in place (returns None); copy() returns a shallow copy. The in operator tests membership in O(n) linear time. For repeated lookups in large sorted lists, use the bisect module for O(log n) binary search.
- list.sort() sorts in place (returns None); sorted() returns a new sorted list without changing the original. Both accept key= (a function applied to each element for comparison) and reverse=True. Python's sort is stable. Use a tuple as the key value to sort by multiple fields simultaneously.
- Nested lists model multi-dimensional data. Access elements with list[row][col]. Create 2D grids with a list comprehension ([[0]*cols for _ in range(rows)]) not list multiplication (which creates aliased inner lists). Slice assignment (list[a:b] = iterable) replaces, inserts, or deletes a range of elements in place.
- List as stack: append() pushes (O(1)), pop() pops from top (O(1)). List as queue: append() enqueues (O(1)), but pop(0) dequeues in O(n). For high-volume FIFO queues, use collections.deque for O(1) popleft(). deque also supports maxlen for a fixed-size sliding window.
- Assignment copies a reference, not the list. To get an independent outer list use copy(), [:], or list() (all shallow copies). Shallow copies share inner mutable objects; modifying a nested list or dict through either copy affects both. Use copy.deepcopy() when nested contents must also be fully independent.

Guided Exercises

Complete the following exercises in VS Code, IDLE, or your preferred Python environment. Each exercise targets a specific concept from this chapter.

1. Write a server fleet manager. Maintain a list of server dicts (host, cpu_pct, status). Implement functions to: add a server (append), promote a server to the front of the priority list (insert at 0), remove a server by host name (iterate and remove), find the most loaded server (max with key=), and print a formatted status table. Test each function and demonstrate that the list is correctly updated after each operation.

2. Implement a multi-key sort demonstrator. Given a list of at least 12 transaction records (id, date, amount, status, region), sort the list five different ways: by amount ascending, by amount descending, by status then amount, by region then date, and by a custom priority (completed first, then pending, then failed — use a mapping dict as the sort key). Print a formatted table for each sort, showing only the first five rows.

3. Build a shift schedule manager using a nested list. Rows represent days (Monday-Friday), columns represent shifts (Morning, Afternoon, Evening). Implement functions to: print the full schedule as an aligned table, look up who is working a specific day and shift, update a shift assignment, find all shifts worked by a specific employee, and swap two employees' assignments for a given day. Test all functions.

4. Implement both a stack and a FIFO queue using the appropriate data structures (list for stack, collections.deque for queue). The stack should support: push, pop, peek, and is_empty. The queue should support: enqueue, dequeue, peek, and is_empty. Both should raise descriptive errors on underflow. Demonstrate both with a realistic use case: the stack as an undo buffer for text edits, the queue as a print job dispatcher.

5. Demonstrate the shallow vs deep copy distinction with a clear experiment. Create a list of configuration dicts (each with nested env_vars and feature_flags sub-dicts). Make three versions: a reference copy (assignment), a shallow copy, and a deep copy. Modify the nested data in each copy and print the original after each modification, showing clearly which types of changes propagate back to the original and which do not. Explain the behaviour in comments.

Challenge Tasks

The following tasks are open-ended. No hints are provided. They require you to design a solution, not just apply a pattern.

1. Design and implement a priority job queue. Each job has an id, priority (1-5, where 5 is highest), payload, status, and attempt_count. The queue should: always process the highest-priority job first (use sorted() with key= rather than keeping the list permanently sorted); support adding, cancelling, and re-queuing failed jobs (up to MAX_ATTEMPTS); produce a status report showing counts by priority and status; and expose a peek() operation that returns

the next job to be processed without removing it. Use a list internally and demonstrate all operations with a realistic sequence of 15 or more job submissions and processing events.

2. Implement a simple in-memory time-series database for sensor readings. The database stores readings as a list of (timestamp, sensor_id, value) tuples. Implement: insert (maintaining sorted order by timestamp using bisect.insort), query by time range returning readings between two timestamps, query by sensor returning all readings for one sensor, rolling average for a sensor over a configurable window size, and anomaly detection that flags readings more than N standard deviations from the rolling mean. Demonstrate with at least 50 simulated readings across at least three sensors.

3. Build a list transformation pipeline. Define a Pipeline class that wraps a list and supports method chaining: filter(predicate), transform(function), sort(key=None, reverse=False), take(n), skip(n), group_by(key_function) (returns a dict of lists), and collect() (returns the resulting list). Each method except collect() should return self (or a new Pipeline) to enable chaining. Demonstrate with a dataset of at least 20 product records, building at least three different report queries using the same pipeline, each using at least four chained operations.

CHAPTER 12 Dictionaries

Think, Code, Create – A Python Programming Journey

A dictionary stores data as key–value pairs, where each key is a unique, immutable label and each value is the data associated with that label. This structure is so pervasive in Python programs that it is hard to write anything non-trivial without one. API responses, configuration files, database records, grouped data, frequency counts, dispatch tables, and cached results are all naturally represented as dictionaries.

This chapter covers the full dictionary toolkit: creation and access, the complete method library, dictionary comprehensions for concise construction, nested dictionaries for hierarchical data, the integration of dicts and lists for tabular and grouped data, iteration and merging patterns including the modern | operator, and strategies for handling missing keys — from get() to defaultdict to dispatch tables.

12.1 Introduction to Dictionaries

A dictionary is created with curly braces containing comma-separated key: value pairs. Keys must be immutable (strings, numbers, tuples of immutables); values may be anything. Since Python 3.7 dictionaries maintain insertion order. Access a value by key with d[key] (raises KeyError if absent) or d.get(key, default) (returns the default if absent). Assign to a new key to add it; assign to an existing key to update it.

Creating and Accessing

Example 1 — Introduction

A server configuration dictionary: access, update, and add key-value pairs.

```
server = {
    "host":        "api-prod-01.internal",
    "port":        8443,
    "protocol":    "HTTPS",
    "timeout_s":   30,
    "max_retries": 3,
    "tags":        ["production", "primary", "region-au"],
}

# Access by key
print(server["host"])            # "api-prod-01.internal"
print(server["port"])            # 8443

# Update an existing key
server["timeout_s"] = 60

# Add a new key
server["tls_version"] = "1.3"
```

```
# Check membership
print("host" in server)         # True
print("password" in server)     # False

# Length
print(f"{len(server)} configuration keys")
```

Mutability and Key Rules

Dictionaries are mutable: keys can be added or removed and values can be replaced. Keys must be hashable — strings, integers, floats, and tuples of hashables are all valid keys. Lists and dicts cannot be keys. If you assign to a key that already exists, the value is replaced; no error is raised. Duplicate keys in a literal are valid syntax but the last assignment wins.

Example 2 — Variation

Demonstrating key mutability, del, and the last-assignment rule for duplicate keys.

```
config = {"retries": 3, "timeout": 30, "debug": False}

# Mutate in place
config["debug"]    = True
config["log_file"] = "/var/log/app.log"
del config["retries"]    # removes the key entirely
print(config)

# Duplicate keys: last one wins (Python processes left-to-right)
service = {"name": "api", "port": 80, "port": 443}
print(service["port"])    # 443

# Valid key types
mixed_keys = {
    "string_key": "value1",
    42:           "value2",
    (1, 2):       "value3",    # tuple of ints is hashable
}

# Invalid: list is not hashable
# bad = {[1, 2]: "value"}   # TypeError: unhashable type: list
```

dict() Constructor and fromkeys()

Dictionaries can also be created with the dict() constructor, accepting keyword arguments or an iterable of key-value pairs. dict.fromkeys(keys, default_value) creates a dict where every key maps to the same initial value — useful for initialising counters, status flags, or lookup tables.

Example 3 — Real-world application

dict() from parallel sequences, and fromkeys() for initialising region counters.

```
# dict() from keyword arguments
db_config = dict(host="db-prod-01", port=5432, dbname="inventory", pool=10)
print(db_config)
```

```
# dict() from list of (key, value) tuples
fields = [("sku", "P-001"), ("qty", 45), ("price", 12.50)]
record = dict(fields)
print(record)

# fromkeys(): initialise all regions with a zero counter
REGIONS = ["AU-NSW", "AU-VIC", "AU-QLD", "NZ", "SG"]
error_counts  = dict.fromkeys(REGIONS, 0)
active_status = dict.fromkeys(REGIONS, True)

error_counts["AU-NSW"] += 3
error_counts["NZ"]     += 1
print("Error counts:", error_counts)
```

12.2 Dictionary Methods

Python dictionaries have twelve built-in methods. The most important ones for everyday use are get(), pop(), update(), setdefault(), keys(), values(), and items(). Understanding when to use each — and the subtle differences between them — is the foundation of effective dictionary programming.

keys(), values(), items()

All three return dynamic view objects, not lists. A view reflects the current state of the dictionary: if you add or remove a key after creating a view, the view updates immediately. Convert to a list with list() if you need a static snapshot. Iterating with items() is by far the most common pattern — it gives both key and value in one step.

Example 1 — Introduction

keys(), values(), and items() views on a service health metrics dictionary.

```
metrics = {
    "api-gateway":  {"cpu": 62, "latency_ms": 45,  "errors": 0},
    "auth-service": {"cpu": 88, "latency_ms": 920, "errors": 7},
    "db-primary":   {"cpu": 35, "latency_ms": 8,   "errors": 0},
}

# keys(): iterate service names
print("Services:", list(metrics.keys()))

# values(): all metric dicts (useful for aggregate calculations)
total_errors = sum(m["errors"] for m in metrics.values())
print(f"Total errors: {total_errors}")

# items(): key + value together (most common pattern)
print("\nService Health:")
for service, m in metrics.items():
    status = "OK" if m["errors"] == 0 and m["cpu"] < 80 else "DEGRADED"
    print(f"  {service:<20} CPU={m['cpu']:>3}%  Latency={m['latency_ms']:>4}ms
[{status}]")
```

get(), pop(), and update()

get(key, default) is the safe accessor: it returns the default (or None if not specified) instead of raising KeyError. pop(key, default) removes and returns a key; with a default argument it is also safe. update() merges another dict or iterable of pairs into the current dict, overwriting existing keys.

Example 2 — Variation

get() for safe access, pop() for extraction, and update() for batch configuration merging.

```
base_config = {
    "host":    "localhost",
    "port":    5432,
    "timeout": 30,
    "debug":   False,
}

# get(): safe access with fallback
pool_size = base_config.get("pool_size", 5)   # key absent -> returns 5
debug     = base_config.get("debug", False)    # key present -> returns False
print(f"pool_size={pool_size}, debug={debug}")

# pop(): remove and return (safe with default)
host = base_config.pop("host")                 # removes "host", returns value
gone = base_config.pop("missing", "fallback")  # absent key -> "fallback"
print(f"Extracted host: {host}")

# update(): merge prod overrides into base config
prod_overrides = {"host": "db-prod-01", "port": 5433, "pool_size": 20}
base_config.update(prod_overrides)
print("Final config:", base_config)
```

setdefault() for Grouped Accumulation

setdefault(key, default) returns the value for key if it exists, or inserts key with the default value and returns that default. It is particularly useful for building grouped data structures where you want to initialise a list or set for a new key on first encounter, without a separate check.

Example 3 — Real-world application

setdefault() grouping transactions by region without an explicit key-existence check.

```
transactions = [
    {"id": "T-001", "region": "AU", "amount": 1_249.00},
    {"id": "T-002", "region": "NZ", "amount":   89.99},
    {"id": "T-003", "region": "AU", "amount":  320.00},
    {"id": "T-004", "region": "SG", "amount": 4_800.00},
    {"id": "T-005", "region": "NZ", "amount":  175.50},
    {"id": "T-006", "region": "AU", "amount":   45.99},
]

# Group transaction IDs by region
by_region = {}
for txn in transactions:
    # setdefault creates an empty list for new regions automatically
    by_region.setdefault(txn["region"], []).append(txn["id"])
```

```python
for region, ids in sorted(by_region.items()):
    print(f"  {region}: {ids}")
```

12.3 Dictionary Comprehensions

A dictionary comprehension {key_expr: value_expr for var in iterable if condition} produces a new dictionary in a single expression. The most practically useful patterns are: inverting a mapping (swapping keys and values), indexing a list of records by a key field, transforming values while keeping keys, and filtering key-value pairs by a condition. Each pattern replaces a for loop and a series of dict assignments with a single readable line.

Inverting and Indexing

Example 1 — Introduction

Inverting a code-to-name mapping; indexing a list of records by a unique key field.

```python
# Invert: code -> name becomes name -> code
status_codes = {200: "OK", 201: "Created", 400: "Bad Request",
                401: "Unauthorized", 404: "Not Found", 500: "Server Error"}

name_to_code = {name: code for code, name in status_codes.items()}
print(name_to_code["Not Found"])   # 404

# Index: list of dicts -> dict keyed by a unique field (O(1) lookup)
products = [
    {"sku": "P-001", "name": "Widget A", "price": 12.50, "stock": 45},
    {"sku": "P-002", "name": "Gadget B", "price": 89.99, "stock": 0},
    {"sku": "P-003", "name": "Module C", "price": 34.95, "stock": 12},
]

by_sku = {p["sku"]: p for p in products}

# Now lookup is O(1) instead of O(n)
item = by_sku.get("P-002")
if item:
    print(f"Found: {item['name']} ${item['price']:.2f}")
```

Transforming and Filtering

Example 2 — Variation

Transform all values in a config dict; filter to only keys matching a condition.

```python
raw_config = {
    "host":         "  db-prod-01  ",
    "port":         " 5432 ",
    "max_pool":     " 20 ",
    "debug":        " false ",
    "log_level":    "  INFO  ",
}

# Transform: strip all string values
cleaned = {k: v.strip() for k, v in raw_config.items()}
```

```
print(cleaned)

# Filter: keep only keys matching a prefix
metrics = {
    "cpu_pct": 72, "mem_pct": 88, "disk_pct": 45,
    "req_per_s": 1_240, "error_rate": 0.02, "latency_ms": 180,
}

resource_metrics = {k: v for k, v in metrics.items() if k.endswith("_pct")}
rate_metrics     = {k: v for k, v in metrics.items() if not
k.endswith("_pct")}

print("Resource:", resource_metrics)
print("Rates:   ", rate_metrics)
```

Building Derived Structures

Example 3 — Real-world application

Building a frequency map and a grouped summary from a transaction list using comprehensions.

```
transactions = [
    {"region": "AU", "status": "completed", "amount": 1_249.00},
    {"region": "NZ", "status": "completed", "amount":    89.99},
    {"region": "AU", "status": "failed",    "amount":  320.00},
    {"region": "AU", "status": "completed", "amount":  450.00},
    {"region": "SG", "status": "completed", "amount": 4_800.00},
    {"region": "NZ", "status": "pending",   "amount":  175.50},
]

# Count completed transactions per region (comprehension + sum)
completed = [t for t in transactions if t["status"] == "completed"]
revenue_by_region = {
    region: sum(t["amount"] for t in completed if t["region"] == region)
    for region in {t["region"] for t in completed}   # unique regions as set
}

for region, total in sorted(revenue_by_region.items()):
    print(f"  {region}: ${total:,.2f}")
```

12.4 Nested Dictionaries

A nested dictionary has dictionaries (or other containers) as its values. This is the natural structure for hierarchical data: a configuration file with sections, an API response with sub-objects, or a database record with grouped fields. Access is chained: d[outer_key][inner_key]. Modification is the same as any dict assignment. The challenge with nested dicts is safe access — accessing a missing inner key raises KeyError. Use .get() chains or the walrus operator to navigate safely.

Creating and Accessing

Example 1 — Introduction

A multi-service configuration registry as a nested dictionary.

```
infrastructure = {
    "api-gateway": {
        "host":     "api.internal",
        "port":     443,
        "replicas": 3,
        "limits":   {"cpu": "500m", "memory": "512Mi"},
    },
    "auth-service": {
        "host":     "auth.internal",
        "port":     8080,
        "replicas": 2,
        "limits":   {"cpu": "250m", "memory": "256Mi"},
    },
    "db-primary": {
        "host":     "db.internal",
        "port":     5432,
        "replicas": 1,
        "limits":   {"cpu": "2000m", "memory": "4Gi"},
    },
}

# Chained access
print(infrastructure["api-gateway"]["port"])              # 443
print(infrastructure["db-primary"]["limits"]["memory"])  # "4Gi"

# Safe access for potentially missing keys
svc = infrastructure.get("cache-layer", {})
cpu_limit = svc.get("limits", {}).get("cpu", "not set")
print(f"cache-layer cpu limit: {cpu_limit}")
```

Iterating and Modifying Nested Dicts

Example 2 — Variation

Iterating a nested registry to produce a formatted report, then applying a batch update.

```
  print(f"Service          Host                      Port   Replicas  CPU Limit")
print("-" * 65)

for service, cfg in infrastructure.items():
    print(f"{service:<16} {cfg['host']:<20} {cfg['port']:>6}",
          f"{cfg['replicas']:>9} {cfg['limits']['cpu']:>10}")

# Batch update: scale all services for a high-load event
for service in infrastructure:
    infrastructure[service]["replicas"] += 1

print(f"\nAfter scaling:")
for svc, cfg in infrastructure.items():
    print(f"  {svc}: {cfg['replicas']} replicas")
```

Converting Between Flat and Nested

Example 3 — Real-world application

Flattening a nested API response dict to a single-level record for CSV export.

```
def flatten_dict(d: dict, prefix: str = "", sep: str = ".") -> dict:
    """
    Flatten a nested dict to a single level.
    {"a": {"b": 1, "c": 2}} -> {"a.b": 1, "a.c": 2}
    """
    result = {}
    for key, value in d.items():
        full_key = f"{prefix}{sep}{key}" if prefix else key
        if isinstance(value, dict):
            result.update(flatten_dict(value, full_key, sep))
        else:
            result[full_key] = value
    return result

api_response = {
    "id":     "SVC-001",
    "status": "running",
    "metrics": {"cpu": 62, "memory": 78},
    "config":  {"replicas": 3, "limits": {"cpu": "500m"}},
}

flat = flatten_dict(api_response)
for key, value in flat.items():
    print(f"  {key:<35} = {value}")
```

12.5 Dictionary and List Integration

The two most common patterns combining dicts and lists are: a list of dicts (tabular data, where each dict is a row) and a dict of lists (grouped data, where each key names a group and its value is the list of items in that group). Both patterns appear constantly in data processing, API consumption, and report generation.

List of Dicts: Tabular Data

A list of dicts is the natural representation of a table: each dict is a row, each key is a column name. This is the structure returned by most database query libraries, CSV parsers, and REST APIs. Operations on this structure — filtering, sorting, grouping, aggregating — are covered thoroughly throughout this chapter.

Example 1 — Introduction

Filtering, sorting, and computing aggregates on a list-of-dict product table.

```
products = [
    {"sku": "P-001", "name": "Widget A", "price": 12.50,  "stock": 45,
"active": True},
    {"sku": "P-002", "name": "Gadget B", "price": 89.99,  "stock": 0,
"active": True},
    {"sku": "P-003", "name": "Module C", "price": 34.95,  "stock": 12,
"active": False},
```

```
    {"sku": "P-004", "name": "Widget D", "price": 8.75,   "stock":
200,"active": True},
    {"sku": "P-005", "name": "Kit E",    "price": 149.00, "stock": 5,
"active": True},
]

# Filter: active products in stock
available = [p for p in products if p["active"] and p["stock"] > 0]

# Sort by price
by_price = sorted(available, key=lambda p: p["price"])

# Aggregate
total_value = sum(p["price"] * p["stock"] for p in available)

print(f"Available: {len(available)} products")
print(f"Inventory value: ${total_value:,.2f}")
print(f"Cheapest: {by_price[0]['name']} ${by_price[0]['price']:.2f}")
print(f"Priciest: {by_price[-1]['name']} ${by_price[-1]['price']:.2f}")
```

Dict of Lists: Grouped Data

A dict of lists groups items by category. The typical construction loop uses setdefault() or defaultdict. Once built, iterating items() gives each group for processing.

Example 2 — Variation

Grouping transactions by region and status using defaultdict, then producing a pivot summary.

```
from collections import defaultdict

transactions = [
    {"id": "T-001", "region": "AU", "status": "completed", "amount":
1_249.00},
    {"id": "T-002", "region": "NZ", "status": "completed", "amount":
89.99},
    {"id": "T-003", "region": "AU", "status": "failed",    "amount":
320.00},
    {"id": "T-004", "region": "SG", "status": "completed", "amount":
4_800.00},
    {"id": "T-005", "region": "AU", "status": "completed", "amount":
450.00},
    {"id": "T-006", "region": "NZ", "status": "pending",   "amount":
175.50},
]

# Group by region
by_region = defaultdict(list)
for txn in transactions:
    by_region[txn["region"]].append(txn)

# Summarise each group
  print(f"Region   Count        Total  Completed")
print("-" * 40)
for region in sorted(by_region):
    group     = by_region[region]
    total     = sum(t["amount"] for t in group)
```

```
    completed = sum(1 for t in group if t["status"] == "completed")
    print(f"{region:<8} {len(group):>6} ${total:>11,.2f} {completed:>10}")
```

Building a Dict Index from a List

Example 3 — Real-world application

Building two indexes from the same list for O(1) lookup by different fields.

```
users = [
    {"id": 1001, "email": "chen.wei@example.com",    "role": "admin",
"active": True},
    {"id": 1002, "email": "amara.o@example.com",     "role": "analyst",
"active": True},
    {"id": 1003, "email": "marcus.s@example.com",    "role": "viewer",
"active": False},
    {"id": 1004, "email": "linh.n@example.com",      "role": "analyst",
"active": True},
]

# Index by ID for fast lookup by numeric ID
by_id    = {u["id"]: u for u in users}

# Index by email for fast lookup by login
by_email = {u["email"]: u for u in users}

# Index by role for group operations
by_role  = defaultdict(list)
for user in users:
    by_role[user["role"]].append(user)

print(by_id[1002]["email"])                       # amara.o@example.com
print(by_email["linh.n@example.com"]["role"]) # analyst
print(f"Analysts: {len(by_role['analyst'])}")
```

12.6 Iteration, Merging, and Dispatch Tables

Three important dictionary techniques that extend the basic iteration patterns: counting and frequency analysis with collections.Counter, merging dictionaries using the | operator and ** unpacking, and using dictionaries as dispatch tables to replace long if-elif chains.

Counter for Frequency Analysis

collections.Counter is a dict subclass designed for counting. Counter(iterable) produces a dict mapping each element to its count. It supports arithmetic operations, most_common(n), and can be updated with more data. It is the idiomatic Python alternative to manually building a frequency dict with a loop.

Example 1 — Introduction

Counter for log-level frequency analysis and finding the most common error messages.

```
from collections import Counter
```

```
log_levels = [
    "INFO", "WARN", "ERROR", "INFO", "INFO", "DEBUG",
    "ERROR", "WARN", "INFO", "CRITICAL", "ERROR", "INFO",
    "DEBUG", "WARN", "ERROR", "INFO",
]

# Counter builds the frequency dict in one call
level_counts = Counter(log_levels)
print("Log level frequencies:")
for level, count in level_counts.most_common():
    bar = "█" * count
    print(f"  {level:<10} {bar:<20} {count:>3}")

# Update Counter with more data
more_logs = ["ERROR", "ERROR", "WARN", "INFO"]
level_counts.update(more_logs)
print(f"\nTotal ERROR events: {level_counts['ERROR']}")

# most_common(n): top N elements
print(f"Top 2 levels: {level_counts.most_common(2)}")
```

Merging Dictionaries

Python 3.9+ provides the | operator to merge two dicts into a new dict, and |= to update in place. Earlier versions use {**d1, **d2} (spread/unpack syntax) which also works in function calls. In all cases, if both dicts share a key the right-hand operand's value wins.

Example 2 — Variation

Config layering: base -> environment -> instance overrides using | and **.

```
# Base configuration (lowest priority)
base = {"host": "localhost", "port": 5432, "timeout": 30,
        "pool_size": 5, "debug": False, "log_level": "INFO"}

# Environment overrides (medium priority)
prod_env = {"host": "db-prod-01", "port": 5433, "pool_size": 20}

# Instance overrides (highest priority)
instance = {"timeout": 60, "log_level": "WARNING"}

# Merge with |: right side wins on conflict (Python 3.9+)
config = base | prod_env | instance
print("Merged config:")
for k, v in config.items():
    print(f"  {k:<12} = {v}")

# ** unpack for Python 3.8 and earlier
config_38 = {**base, **prod_env, **instance}
assert config == config_38   # identical result

# |= for in-place update (equivalent to dict.update)
base |= prod_env
print(f"\nbase host is now: {base['host']}")
```

Dispatch Tables

A dispatch table is a dict mapping keys to functions (or callables). It replaces a long if-elif chain with a single dict lookup — faster, easier to extend, and easier to read. Add a new case by inserting a key-value pair; no elif needed. Combine with .get(key, default_function) to handle unknown commands gracefully.

Example 3 — Real-world application

A command dispatcher for a deployment CLI: dict maps command names to handler functions.

```
def cmd_deploy(args):    print(f"Deploying {args.get('service')} to
{args.get('env')}")
def cmd_rollback(args):  print(f"Rolling back {args.get('service')}")
def cmd_status(args):    print(f"Checking status of {args.get('service')}")
def cmd_scale(args):     print(f"Scaling {args.get('service')} to
{args.get('replicas')} replicas")
def cmd_logs(args):      print(f"Streaming logs for {args.get('service')}")
def cmd_unknown(args):   print(f"Unknown command. Valid: {sorted(COMMANDS)}")

# Dispatch table: command name -> handler function
COMMANDS = {
    "deploy":   cmd_deploy,
    "rollback": cmd_rollback,
    "status":   cmd_status,
    "scale":    cmd_scale,
    "logs":     cmd_logs,
}

# Execute a command: dict lookup replaces an if-elif chain
def execute(command: str, args: dict):
    handler = COMMANDS.get(command, cmd_unknown)
    handler(args)

execute("deploy",  {"service": "api-gateway", "env": "prod"})
execute("scale",   {"service": "worker",       "replicas": 5})
execute("restart", {})
```

12.7 Handling Missing Keys and Default Values

Every real-world dict operation must consider the possibility that a requested key is absent. Python provides a graduated set of strategies: bracket access for when absence is a bug; get() for safe reads; setdefault() for initialisation-on-first-access; defaultdict for entire dicts where absence should silently produce a default; and try/except for cases where the error handling logic is complex or needs to log the failure.

get() and Chained get()

Example 1 — Introduction

get() with defaults and chained get() for navigating potentially incomplete nested dicts.

```
api_responses = [
    {"status": 200, "data": {"user": {"id": 1001, "role": "admin"}}},
    {"status": 200, "data": {"user": {"id": 1002}}},         # no role
    {"status": 200, "data": {}},                              # no user
    {"status": 404, "error": "Not found"},                    # no data
```

```
]

for resp in api_responses:
    # Chained get() -- each level falls back to {} if the key is absent
    user = resp.get("data", {}).get("user", {})
    user_id = user.get("id", "unknown")
    role    = user.get("role", "viewer")   # default role if not specified
    print(f"  [{resp['status']}] id={user_id}, role={role}")
```

defaultdict for Automatic Default Values

defaultdict(factory) creates a dict that calls factory() whenever a missing key is accessed, inserting the result as the key's value. Common factories: list for grouping, int for counting (0), set for unique-element grouping, and lambda expressions for complex defaults. The factory is called only on missing keys; existing keys behave identically to a regular dict.

Example 2 — Variation

defaultdict(list) for grouping, defaultdict(int) for counting, and defaultdict(set) for unique tags.

```
from collections import defaultdict

events = [
    {"service": "api",   "type": "request",  "tag": "v2"},
    {"service": "auth",  "type": "error",    "tag": "v1"},
    {"service": "api",   "type": "error",    "tag": "v2"},
    {"service": "cache", "type": "request",  "tag": "v2"},
    {"service": "api",   "type": "request",  "tag": "v1"},
]

# defaultdict(list): group event types per service
by_service = defaultdict(list)
for e in events:
    by_service[e["service"]].append(e["type"])

# defaultdict(int): count events per type
type_counts = defaultdict(int)
for e in events:
    type_counts[e["type"]] += 1

# defaultdict(set): unique tags seen per service
tags_per_service = defaultdict(set)
for e in events:
    tags_per_service[e["service"]].add(e["tag"])

for svc, types in sorted(by_service.items()):
    tags = tags_per_service[svc]
    print(f"  {svc:<8} events={len(types)} tags={tags}")
```

try/except for Complex Error Handling

When a missing key should trigger logging, an alert, or a multi-step fallback rather than just returning a default value, try/except KeyError gives you the flexibility to handle the error with full logic. This is also the correct pattern when you are accessing a chain of nested dicts and want to catch KeyError at any level in a single handler.

Example 3 — Real-world application

try/except KeyError when fetching a config section that may be entirely absent, with logging.

```
import sys

full_config = {
    "database": {"host": "db-prod-01", "port": 5432},
    "cache":    {"host": "cache-01",   "port": 6379},
    # "queue" section is missing
}

REQUIRED_SECTIONS = ["database", "cache", "queue", "storage"]

for section in REQUIRED_SECTIONS:
    try:
        cfg = full_config[section]
        host = cfg["host"]   # also raises KeyError if "host" is absent
        print(f"  [{section}] host={host}")
    except KeyError as e:
        print(f"  [{section}] MISSING key: {e}", file=sys.stderr)
```

Chapter 12 Summary

- Dictionaries store key–value pairs in insertion order (Python 3.7+). Keys must be immutable (strings, numbers, tuples of immutables); values may be any type. Create with {}, dict(), or dict.fromkeys(keys, default). Access with d[key] (KeyError if absent) or d.get(key, default). Assign to add or update; del d[key] removes.
- Core methods: keys() / values() / items() return live views (not lists). get(k, default) safe-reads. pop(k, default) removes and returns. update(other) merges. setdefault(k, default) returns existing value or inserts and returns default. All mutation methods return None (not a new dict).
- Dictionary comprehensions {k: v for ... if ...} support inverting mappings, indexing lists of records by a key field, transforming values, and filtering entries. Combine with set literals in the for clause to iterate unique values without importing anything.
- Nested dicts model hierarchical data. Access is chained: d[outer][inner]. Use chained .get() for safe navigation: d.get("outer", {}).get("inner", default). The flatten_dict() pattern converts nested dicts to dot-notation flat dicts for export or logging.
- List-of-dicts (tabular) and dict-of-lists (grouped) are the two fundamental combined patterns. Build indexes (by_sku, by_email) with dict comprehensions for O(1) lookup. Build grouped indexes with defaultdict(list) or setdefault().
- Counter(iterable) counts frequencies and supports most_common(n) and arithmetic. Merge dicts with d1 | d2 (Python 3.9+, right side wins) or {**d1, **d2} (older). Dispatch tables (dict mapping string commands to functions) replace if-elif chains and are more extensible.
- Missing-key strategies: get(k, default) for simple fallback; setdefault() for initialisation-on-first-access; defaultdict(factory) for entire dicts where every missing key should produce a default; try/except KeyError for cases requiring complex error logic or multi-level nested access.

Guided Exercises

Complete the following exercises in VS Code, IDLE, or your preferred Python environment. Each exercise targets a specific concept from this chapter.

1. Build a service configuration registry. Store at least five services as a nested dict where each service has host, port, replicas, and limits (itself a dict with cpu and memory keys). Write functions to: get a service's CPU limit safely using chained .get(); add a new service; update an existing service's replica count; and produce a formatted summary table. Test all operations including access to a service that does not exist.

2. Write a transaction processor using dict comprehensions. Given a list of transaction dicts (id, region, status, amount, currency), produce: (1) an index by transaction id; (2) a dict of total amounts by region for completed transactions only; (3) a dict of status -> count using Counter; (4) a filtered dict keeping only transactions above a threshold amount. Print all four results.

3. Implement a multi-level configuration merge system. Define a base config dict, an environment config, and an instance config, each potentially having nested keys. Write a deep_merge(base, override) function that merges dicts recursively (nested dicts are merged, not replaced) rather than using the flat | operator. Demonstrate that deep_merge preserves unchanged inner keys while overrides take effect, and test with at least three levels of nesting.

4. Build a log analyser using defaultdict and Counter. Given a list of log records (each a dict with timestamp, level, service, and message), produce: counts by level using Counter; grouped records by service using defaultdict(list); a set of unique services using defaultdict(set); and the top 3 most common error messages. Print a formatted report with all four analyses.

5. Design and implement a dispatch table based command runner. Define at least eight command handler functions for a hypothetical infrastructure CLI (e.g. deploy, rollback, scale, restart, logs, status, health, config). Build a COMMANDS dispatch dict. Write an execute(command, args) function that looks up and calls the correct handler, or prints a "command not found" message with suggestions if the input is not recognised. Demonstrate with at least six different command invocations.

Challenge Tasks

The following tasks are open-ended. No hints are provided. They require you to design a solution, not just apply a pattern.

1. Build a configurable data pivot tool. Given a list of records (at least 20 dicts), a row_key field name, a col_key field name, and an aggregate function (sum, mean, count, min, max), produce a pivot table as a nested dict: result[row_value][col_value] = aggregated_value. Print the pivot table with row and column headers aligned. Support at least: pivoting transaction data by region and status (aggregate: sum of amounts), and pivoting server metrics by service and metric type (aggregate: mean). The pivot logic must work for any dataset with the specified fields.

---

2. Implement a simple in-memory key-value store with TTL (time-to-live) support. Internally store values as {"value": v, "expires_at": timestamp or None}. Implement: set(key, value, ttl_seconds=None), get(key, default=None) (returns default if key absent or expired), delete(key), exists(key), keys() (only non-expired keys), and expire_all() (removes all expired entries). Demonstrate with a realistic caching scenario: cache API responses with a 5-second TTL, verify that expired entries are not returned, and show that unexpired entries survive past shorter TTLs.

---

3. Design and implement a schema-validated dictionary type. Define a schema as a dict mapping field names to type descriptors (required: bool, type: type or tuple of types, default: optional, validator: optional callable). Write a validate(data, schema) function that: checks all required fields are present, checks each value's type matches the schema, applies any custom validators, collects all errors (not just the first), and returns (is_valid: bool, errors: list). Demonstrate with at least three different schemas (e.g. a product record, a server config, a user account) and test each with valid data and at least two different invalid inputs.

CHAPTER 13 Tuples and Sets

Think, Code, Create – A Python Programming Journey

Tuples and sets solve two distinct problems that lists and dicts cannot. A tuple is an immutable ordered sequence — once created it cannot change, which makes it safe to use as a dictionary key, guarantees that data passed between functions remains intact, and communicates intent: this collection should not be modified. A set is an unordered collection of unique hashable elements with O(1) membership testing — it is the right structure whenever uniqueness, fast lookup, or mathematical set relationships (union, intersection, difference) are the primary concern.

This chapter covers tuples completely: packing, unpacking, extended unpacking, named tuples, and their use as dictionary keys and function return values. It then covers sets: creation, the full method library including discard vs remove, frozenset, all four set operations with both operator and method syntax, in-place update operators, and practical applications in permission systems, data change detection, and tag filtering.

13.1 Tuples

A tuple is an immutable ordered sequence. It is created with parentheses (optional in most contexts) and comma-separated values. A single-element tuple requires a trailing comma: (42,) is a tuple; (42) is just an integer in parentheses. Tuples support all read operations that lists support — indexing, slicing, len(), in, iteration, and index()/count() — but not any mutation.

Creating, Indexing, and Immutability

Example 1 — Introduction

Tuple creation, indexing, and the TypeError raised by any attempted mutation.

```
# Tuple literal (parentheses optional when context is unambiguous)
connection = ("db-prod-01", 5432, "inventory", "app_user")

# Indexing and slicing
host = connection[0]          # "db-prod-01"
port = connection[1]          # 5432
creds = connection[2:]        # ("inventory", "app_user")

# Membership and length
print(5432 in connection)     # True
print(len(connection))        # 4

# Immutability: tuples cannot be mutated
try:
    connection[1] = 5433
except TypeError as e:
    print(f"TypeError: {e}")
```

```
# Single-element tuple requires trailing comma
single = ("db-prod-01",)   # tuple
not_a_tuple = ("db-prod-01") # just a string
print(type(single))            # <class 'tuple'>
print(type(not_a_tuple))       # <class 'str'>
```

Packing, Unpacking, and Extended Unpacking

Tuple packing is assigning multiple values to a single name: coords = 51.5, -0.1 packs two floats into a tuple without parentheses. Unpacking assigns the elements of a tuple to multiple names in one statement. The * operator in an unpacking expression collects "the rest" into a list, enabling flexible first/last extraction.

Example 2 — Variation

*Tuple packing and unpacking; extended unpacking with * for parsing structured records.*

```
# Packing: parentheses are optional
server_record = "api-prod-01", "10.0.1.10", 443, "healthy"

# Basic unpacking
hostname, ip, port, status = server_record
print(f"{hostname} ({ip}:{port}) -- {status}")

# Extended unpacking: * collects the middle portion
log_fields = ("2024-07-15", "09:14:02", "ERROR", "auth", "login failed",
"attempt 3")
date, time, level, *details = log_fields
print(f"Level: {level}, Details: {details}")

# First and last with *
first, *middle, last = [10, 20, 30, 40, 50]
print(f"first={first}, middle={middle}, last={last}")

# Swap two variables using tuple unpacking (no temp variable needed)
a, b = 100, 200
a, b = b, a
print(f"a={a}, b={b}")   # a=200, b=100
```

Tuples as Dictionary Keys and in Sorting

Because tuples are hashable (assuming all their elements are hashable), they can be used as dictionary keys. A common pattern is using a tuple of two or more values as a composite key. Tuples are also the standard return value for sorted() with a key function: return a tuple to sort by multiple fields simultaneously.

Example 3 — Real-world application

Composite tuple keys for a 2D grid lookup; tuple sorting key for multi-field ordering.

```
# Composite (row, col) keys in a sparse grid
grid_values = {
    (0, 0): "switch-core-01",
    (0, 1): "switch-core-02",
```

```
    (1, 0): "server-rack-A",
    (1, 2): "server-rack-C",
}
print(grid_values.get((1, 0), "empty"))    # "server-rack-A"
print(grid_values.get((1, 1), "empty"))    # "empty"

# Multi-field tuple sort key
incidents = [
    {"id": "INC-003", "severity": 2, "service": "auth"},
    {"id": "INC-001", "severity": 1, "service": "api"},
    {"id": "INC-004", "severity": 1, "service": "auth"},
    {"id": "INC-002", "severity": 2, "service": "cache"},
]
# Sort by severity (ascending), then service name, then id
ranked = sorted(incidents, key=lambda i: (i["severity"], i["service"],
i["id"]))
for inc in ranked:
    print(f"  [{inc['severity']}] {inc['id']}  {inc['service']}")
```

13.2 When to Use Tuples

The choice between a tuple and a list is fundamentally about intent and correctness. Use a tuple when the number of elements is fixed and each position has a defined meaning — a 2D coordinate, a database connection parameter set, a function return value containing a result and a status. Use a list when the collection may grow or shrink, or when the elements are interchangeable rather than positionally significant.

Named Tuples for Readable Records

collections.namedtuple creates a tuple subclass whose fields are accessible by name as well as by index. This is the right choice when a tuple represents a structured record — it gives the positional benefits of a tuple (immutable, hashable, light memory footprint) with the readability of field names. Python 3.6+ also provides typing.NamedTuple for the same purpose with type annotations.

Example 1 — Introduction

namedtuple for readable server and network event records.

```
from collections import namedtuple

# Define a named tuple type
ServerRecord = namedtuple("ServerRecord", ["host", "ip", "port", "status"])

# Create instances
servers = [
    ServerRecord("api-prod-01", "10.0.1.10", 443,  "healthy"),
    ServerRecord("db-primary",  "10.0.1.20", 5432, "healthy"),
    ServerRecord("cache-01",    "10.0.1.30", 6379, "degraded"),
]

# Access by name (readable) or by index (compatible with plain tuple usage)
for s in servers:
    print(f"  {s.host:<15} {s.ip:<14} :{s.port:<5} {s.status}")
```

```
# Still a tuple: hashable, unpackable, works as dict key
host, ip, port, status = servers[0]
lookup = {servers[0]: "primary endpoint"}
print(f"\nLookup: {lookup[servers[0]]}")
```

Returning Multiple Values

Python functions can return multiple values as a tuple — the parentheses are optional. The caller can either receive the tuple as one name or unpack it into multiple names. This is cleaner than returning a list (which implies the count may vary) or a dict (which has more overhead). Named tuples work particularly well here because they give field names to return values.

Example 2 — Variation

A function returning a named tuple for a parse result with success flag and extracted fields.

```
from collections import namedtuple

ParseResult = namedtuple("ParseResult", ["ok", "timestamp", "level",
"message", "raw"])

def parse_log_line(line: str) -> ParseResult:
    """
    Parse a log line. Returns a ParseResult namedtuple.
    ok=False if the line cannot be parsed.
    """
    parts = line.strip().split(None, 3)
    if len(parts) < 3:
        return ParseResult(False, None, None, None, line)
    ts, level, *rest = parts
    return ParseResult(
        ok        = True,
        timestamp = ts,
        level     = level.upper(),
        message = rest[0] if rest else "",
        raw       = line.strip(),
    )

lines = ["2024-07-15T09:14 ERROR disk quota exceeded", "bad line", ""]
for line in lines:
    r = parse_log_line(line)
    if r.ok:
        print(f"  [{r.level}] {r.timestamp}: {r.message}")
    else:
        print(f"  [SKIP] Could not parse: {r.raw!r}")
```

13.3 Introduction to Sets

A set is an unordered collection of unique hashable elements. "Unordered" means there is no guaranteed iteration order and elements cannot be accessed by index. "Unique" means each value appears at most once — adding a duplicate silently has no effect. "Hashable" means the elements must be immutable (strings, numbers, tuples of immutables) — lists and dicts cannot be set elements.

Sets are created with curly braces {elem, elem, ...} or set(iterable). An empty set must be created with set() because {} creates an empty dict. frozenset is the immutable variant of set — it cannot be mutated, which makes it hashable and usable as a dictionary key or as an element of another set.

Creating Sets and the frozenset

Example 1 — Introduction

Creating sets from literals and iterables; frozenset for immutable use as a dict key.

```
# Set from a literal (duplicates silently removed)
approved_ips = {"10.0.1.10", "10.0.1.20", "10.0.1.30", "10.0.1.10"}
print(f"Approved IPs: {len(approved_ips)} entries")  # 3, not 4

# Set from an iterable (useful for deduplication)
log_levels = ["INFO", "WARN", "ERROR", "INFO", "DEBUG", "ERROR", "WARN"]
unique_levels = set(log_levels)
print(f"Unique levels: {sorted(unique_levels)}")

# Empty set: must use set(), not {}
seen_ids  = set()       # empty set
empty_dict = {}          # empty dict (NOT a set)
print(type(seen_ids), type(empty_dict))

# frozenset: immutable, hashable -- can be a dict key or set element
REQUIRED_HEADERS = frozenset({"Content-Type", "Authorization", "X-Request-
ID"})
OPTIONAL_HEADERS = frozenset({"X-Correlation-ID", "X-User-Agent"})

# frozenset as a dict key
endpoint_policies = {REQUIRED_HEADERS: "strict", OPTIONAL_HEADERS: "relaxed"}
print(type(REQUIRED_HEADERS))  # <class 'frozenset'>
```

Adding, Removing, and Discarding

add(elem) adds a single element (no-op if already present). remove(elem) removes an element and raises KeyError if absent. discard(elem) removes an element but silently ignores absence — the safe alternative to remove() when the element may or may not be present. pop() removes and returns an arbitrary element (since sets are unordered, there is no "first" element). clear() empties the set.

Example 2 — Variation

add(), discard(), and remove() managing an active session set.

```
active_sessions = {"sess-001", "sess-002", "sess-003"}

# add(): no-op if already present
active_sessions.add("sess-004")
active_sessions.add("sess-001")  # already present, silently ignored
print(f"Active: {len(active_sessions)} sessions")

# discard(): safe removal -- no error if element absent
active_sessions.discard("sess-999")  # absent -- no error
active_sessions.discard("sess-002")  # present -- removed
```

```
# remove(): raises KeyError if absent (use when absence is a bug)
try:
    active_sessions.remove("sess-999")
except KeyError:
    print("Session not found (expected)")

# Membership test: O(1)
print("sess-001" in active_sessions)   # True
print("sess-002" in active_sessions)   # False (discarded)
```

13.4 Set Operations

Sets support the four standard mathematical operations as both operators and methods. The operator forms require both operands to be sets. The method forms (union(), intersection(), etc.) accept any iterable as the argument, making them more flexible when one operand is a list or other non-set collection. In-place variants (|=, &=, -=, ^=) modify the set in place and return None.

Union, Intersection, Difference, Symmetric Difference

Example 1 — Introduction

All four operations applied to permission sets in a role-based access control context.

```
# Permission sets for two roles
analyst_perms  = {"reports.read", "dashboards.read", "data.export",
                  "alerts.read"}
manager_perms  = {"reports.read", "reports.write", "dashboards.read",
                  "dashboards.write", "users.read", "billing.read"}

# Union (|): everything either role can do
combined = analyst_perms | manager_perms
print(f"Combined permissions: {len(combined)}")

# Intersection (&): permissions both roles share
shared = analyst_perms & manager_perms
print(f"Shared: {sorted(shared)}")

# Difference (-): what analysts have that managers don't
analyst_only = analyst_perms - manager_perms
print(f"Analyst-only: {sorted(analyst_only)}")

# Symmetric difference (^): permissions unique to each role
unique_to_each = analyst_perms ^ manager_perms
print(f"Unique to each: {sorted(unique_to_each)}")
```

Subset, Superset, and Disjoint

issubset(other) returns True if every element of the set is in other. issuperset(other) returns True if the set contains every element of other. isdisjoint(other) returns True if the two sets share no elements. These are useful for checking permission containment, required-field presence, and compatibility between configurations.

Example 2 — Variation

Subset/superset/disjoint for permission containment and required-field validation.

```
ADMIN_PERMS  = {"read", "write", "delete", "manage_users", "billing"}
EDITOR_PERMS = {"read", "write"}
VIEWER_PERMS = {"read"}

# Is viewer a subset of editor?
print(VIEWER_PERMS.issubset(EDITOR_PERMS))   # True
print(EDITOR_PERMS.issuperset(VIEWER_PERMS)) # True

# Do these two feature sets overlap?
feature_set_a = {"logging", "metrics", "tracing"}
feature_set_b = {"caching", "rate_limiting", "auth"}
print(feature_set_a.isdisjoint(feature_set_b))  # True (no overlap)

# Validate that a request includes all required fields
REQUIRED_FIELDS = frozenset({"customer_id", "amount", "currency",
"timestamp"})
request_fields  = {"customer_id", "amount", "currency", "timestamp",
"reference"}

missing = REQUIRED_FIELDS - request_fields
if not missing:
    print("All required fields present")
else:
    print(f"Missing: {missing}")
```

In-Place Operations and Method vs Operator

Example 3 — Real-world application

In-place operators updating a permission set; method forms accepting a non-set iterable.

```
# In-place operators: modify the set in place, return None
user_perms = {"read", "write"}

user_perms |= {"delete", "export"}   # add permissions
print(f"After grant:  {sorted(user_perms)}")

user_perms -= {"delete"}             # revoke permission
print(f"After revoke: {sorted(user_perms)}")

user_perms &= {"read", "export"}     # keep only these
print(f"After filter: {sorted(user_perms)}")

# Method forms accept any iterable (not just sets)
granted_from_api = ["billing", "reports"]  # a list, not a set

user_perms = user_perms.union(granted_from_api)       # works
# user_perms = user_perms | granted_from_api          # TypeError!

user_perms = user_perms.difference(["reports"])        # works
print(f"Final: {sorted(user_perms)}")
```

13.5 O(1) Membership Testing and Deduplication

The two most practically important properties of sets are constant-time membership testing and automatic deduplication. Both arise from the underlying hash-table implementation. Understanding when to convert a list to a set — and the trade-offs involved — enables significant performance improvements in lookup-heavy code.

Membership Testing: O(1) vs O(n)

Checking whether an element is in a list requires a linear scan: in the worst case, every element is examined. Checking in a set computes the element's hash and checks one bucket: this is O(1) regardless of set size. For repeated lookups over a large collection, converting the collection to a set before the lookup loop is a straightforward optimisation.

Example 1 — Introduction

Converting a large blocklist to a set for O(1) lookup during request processing.

```
import time

# Simulated blocklist of 100,000 entries
blocked_ips_list = [f"192.168.{i//256}.{i%256}" for i in range(100_000)]
blocked_ips_set  = set(blocked_ips_list)

test_ips = ["10.0.0.1", "192.168.100.50", "172.16.0.1"] * 10_000

# List lookup: O(n) per check
start = time.perf_counter()
list_hits = sum(1 for ip in test_ips if ip in blocked_ips_list)
list_time = time.perf_counter() - start

# Set lookup: O(1) per check
start = time.perf_counter()
set_hits  = sum(1 for ip in test_ips if ip in blocked_ips_set)
set_time  = time.perf_counter() - start

print(f"List lookup: {list_time:.3f}s  ({list_hits} hits)")
print(f"Set lookup:  {set_time:.3f}s   ({set_hits} hits)")
print(f"Speedup: ~{list_time/set_time:.0f}x")
```

Deduplication Patterns

Converting a sequence to a set removes all duplicates in O(n) time. Converting back to a sorted list with sorted() gives a deduplicated, ordered result. When order of first appearance must be preserved, use a different pattern: iterate and track seen elements with a set.

Example 2 — Variation

Three deduplication patterns: set conversion, order-preserving, and counting duplicates.

```
# Pattern 1: unordered deduplication (simplest)
log_sources = ["api-01", "db-01", "api-01", "cache-01", "api-01", "db-01"]
unique_sources = sorted(set(log_sources))
print(f"Unique sources: {unique_sources}")

# Pattern 2: order-preserving deduplication
```

```
def deduplicate_ordered(items):",
    seen = set()
    return [x for x in items if not (x in seen or seen.add(x))]

events = ["login", "request", "login", "error", "request", "logout"]
print(f"First-seen order: {deduplicate_ordered(events)}")

# Pattern 3: identify AND count duplicates
from collections import Counter
skus = ["P-001", "P-002", "P-001", "P-003", "P-002", "P-001"]
counts  = Counter(skus)
dupes   = {sku for sku, n in counts.items() if n > 1}
print(f"Duplicate SKUs: {dupes}")
print(f"Duplicate counts: {dict(counts)}")
```

13.6 Sets in Real-World Applications

The combination of O(1) membership testing, automatic deduplication, and set algebra makes sets the right tool for a recurring class of problems: access control and permission management, detecting changes between two snapshots of data, tag-based filtering, and computing coverage or overlap between datasets.

Permission and Access Control

Example 1 — Introduction

A role-permission system using set operations to compute effective permissions and validate access.

```
ROLE_PERMISSIONS = {
    "viewer":   frozenset({"read"}),
    "analyst":  frozenset({"read", "export", "report"}),
    "editor":   frozenset({"read", "write", "delete"}),
    "admin":    frozenset({"read", "write", "delete", "manage", "billing"}),
}

def effective_permissions(roles: list[str]) -> frozenset:
    """Union of all permissions across a user's roles."""
    perms = set()
    for role in roles:
        perms |= ROLE_PERMISSIONS.get(role, set())
    return frozenset(perms)

def can_perform(user_roles: list[str], required_permission: str) -> bool:
    return required_permission in effective_permissions(user_roles)

users = [
    ("Chen Wei",   ["analyst", "editor"]),
    ("Amara Okafor", ["viewer"]),
    ("Marcus Silva", ["admin"]),
]

for name, roles in users:
    perms = effective_permissions(roles)
    can_bill = can_perform(roles, "billing")
    print(f"  {name:<16} perms={len(perms)}  billing={can_bill}")
```

Change Detection Between Snapshots

Example 2 — Variation

Detecting added, removed, and common elements between two snapshots of a service registry.

```
# Service registry snapshots taken 5 minutes apart
snapshot_before = {
    "api-gateway", "auth-service", "db-primary", "cache-01",
    "worker-01",   "worker-02",
}
snapshot_after = {
    "api-gateway", "auth-service", "db-primary", "cache-01",
    "worker-01",   "worker-03",   "metrics-collector",
}

# Set operations reveal exactly what changed
added   = snapshot_after  - snapshot_before
removed = snapshot_before - snapshot_after
stable  = snapshot_before & snapshot_after

print(f"Added   ({len(added)}):   {sorted(added)}")
print(f"Removed ({len(removed)}): {sorted(removed)}")
print(f"Stable  ({len(stable)}):  {sorted(stable)}")

# Flag: any critical services removed?
CRITICAL = frozenset({"db-primary", "auth-service", "api-gateway"})
critical_removed = removed & CRITICAL
if critical_removed:
    print(f"ALERT: critical services lost: {critical_removed}")
```

Tag-Based Filtering

Example 3 — Real-world application

Filtering a resource catalogue using tag set operations: AND (intersection) and OR (union) semantics.

```
resources = [
    {"name": "api-prod-01",  "tags": {"production", "api",     "primary",
"au"}},
    {"name": "api-staging",  "tags": {"staging",    "api",     "au"}},
    {"name": "db-prod",      "tags": {"production", "database","primary",
"au"}},
    {"name": "db-replica",   "tags": {"production", "database","replica",
"sg"}},
    {"name": "cache-prod",   "tags": {"production", "cache",   "au"}},
]

def filter_resources(resources, require_all=None, require_any=None,
exclude=None):
    """
    Filter resources by tag sets.
    require_all: resource must have ALL these tags
    require_any: resource must have AT LEAST ONE of these tags
    exclude:     resource must have NONE of these tags
    """
```

```
    require_all = set(require_all or [])
    require_any = set(require_any or [])
    exclude     = set(exclude or [])

    return [
        r for r in resources
        if (not require_all or require_all.issubset(r["tags"]))
        and (not require_any or not require_any.isdisjoint(r["tags"]))
        and r["tags"].isdisjoint(exclude)
    ]

# AU production resources that are not replicas
results = filter_resources(resources, require_all={"production", "au"},
exclude={"replica"})
for r in results:
    print(f"  {r['name']}: {r['tags']}")
```

13.7 Set Comprehensions

A set comprehension {expression for variable in iterable if condition} creates a set in a single expression, with automatic deduplication. Like list comprehensions, they are concise and readable for filter-and-transform patterns. Unlike list comprehensions, the result is unordered and contains only unique values, which makes them the right tool when uniqueness is the goal and order is irrelevant.

Basic Set Comprehensions

Example 1 — Introduction

Set comprehensions to extract unique values and normalise a field across a collection.

```
transactions = [
    {"id": "T-001", "region": "AU",  "status": "completed", "currency":
"AUD"},
    {"id": "T-002", "region": "NZ",  "status": "pending",   "currency":
"NZD"},
    {"id": "T-003", "region": "AU",  "status": "failed",    "currency":
"AUD"},
    {"id": "T-004", "region": "SG",  "status": "completed", "currency":
"SGD"},
    {"id": "T-005", "region": "AU",  "status": "completed", "currency":
"AUD"},
]

# Unique regions (order not important)
active_regions = {t["region"] for t in transactions if t["status"] ==
"completed"}
print(f"Active regions: {sorted(active_regions)}")

# Unique currencies across all transactions
currencies = {t["currency"] for t in transactions}
print(f"Currencies in use: {sorted(currencies)}")

# Unique statuses normalised to lowercase
statuses = {t["status"].lower() for t in transactions}
```

```
print(f"Status values: {sorted(statuses)}")
```

Comprehensions with Conditions and Transformations

Example 2 — Variation

Set comprehension to extract unique error codes from log records with filtering.

```
log_records = [
    {"level": "ERROR", "code": "E001", "service": "auth"},
    {"level": "INFO",  "code": None,   "service": "api"},
    {"level": "ERROR", "code": "E002", "service": "db"},
    {"level": "WARN",  "code": "W001", "service": "cache"},
    {"level": "ERROR", "code": "E001", "service": "auth"},  # duplicate
    {"level": "ERROR", "code": "E003", "service": "api"},
    {"level": "ERROR", "code": None,   "service": "api"},   # code missing
]

# Unique error codes (excluding None values)
error_codes = {r["code"] for r in log_records
               if r["level"] == "ERROR" and r["code"] is not None}
print(f"Distinct error codes: {sorted(error_codes)}")

# Services that produced errors (unique)
error_services = {r["service"] for r in log_records if r["level"] == "ERROR"}
all_services    = {r["service"] for r in log_records}
clean_services  = all_services - error_services

print(f"Error services:  {sorted(error_services)}")
print(f"Clean services:  {sorted(clean_services)}")
```

frozenset Comprehensions and Nested Sets

Example 3 — Real-world application

Building a frozenset of required tags for each environment using a comprehension.

```
# Environment tag requirements (each value is a frozenset)
ENV_TAG_REQUIREMENTS = {
    env: frozenset({f"env:{env}", "monitored", "backup"}
                   | ({"ha-enabled"} if env == "prod" else set()))
    for env in ["dev", "staging", "prod"]
}

for env, required in sorted(ENV_TAG_REQUIREMENTS.items()):
    print(f"  {env:<8}: {sorted(required)}")

# Check a resource against its environment requirements
resource = {"name": "api-prod-01", "env": "prod",
            "tags": frozenset({"env:prod", "monitored", "backup", "ha-
enabled", "au"})}

required = ENV_TAG_REQUIREMENTS[resource["env"]]
missing  = required - resource["tags"]
```

```
if not missing:
    print(f"\n{resource['name']}: all required tags present")
else:
    print(f"\n{resource['name']}: missing tags {missing}")
```

Chapter 13 Summary

- Tuples are immutable ordered sequences. A single-element tuple requires a trailing comma: (x,). Tuples support all read operations (indexing, slicing, in, len, iteration, index(), count()) but not mutation. They are hashable when all elements are hashable, making them valid dict keys and set elements.
- Tuple unpacking assigns elements to multiple names: a, b, c = my_tuple. Extended unpacking with * collects the remaining elements into a list: first, *middle, last = seq. namedtuple creates a tuple subclass with named fields, combining immutability and readability for structured records and function return values.
- Sets are unordered collections of unique hashable elements. Create with {a, b, c} or set(iterable). Empty set must be set(), not {} (which creates a dict). add() inserts; remove() deletes (KeyError if absent); discard() deletes silently; pop() removes an arbitrary element. frozenset is the immutable variant, usable as a dict key.
- Four set operations: | (union, all elements from both), & (intersection, elements common to both), - (difference, in left but not right), ^ (symmetric difference, unique to each). In-place variants: |=, &=, -=, ^=. Operator forms require both operands to be sets; method forms (union(), intersection(), etc.) accept any iterable.
- issubset(other), issuperset(other), and isdisjoint(other) test containment and overlap relationships. These are the correct tools for permission containment checks, required-field validation, and compatibility testing.
- Set membership testing is O(1) (hash lookup) vs O(n) for lists. Convert a large lookup collection to a set before a repeated-check loop to obtain significant speedups. set(collection) removes duplicates in O(n). Order-preserving deduplication requires a separate seen set: [x for x in items if not (x in seen or seen.add(x))].
- Set comprehensions {expr for var in iterable if condition} produce sets with automatic deduplication. Use them to extract unique field values, normalise and deduplicate strings, and build environment or tag requirement sets. frozenset values in comprehension results can serve as dict keys.

Guided Exercises

Complete the following exercises in VS Code, IDLE, or your preferred Python environment. Each exercise targets a specific concept from this chapter.

1. Define a namedtuple NetworkEvent with fields: timestamp, source_ip, dest_ip, protocol, port, and action ("allow"/"block"). Create a list of at least ten NetworkEvent instances. Write functions to: filter events by protocol; find all unique source IPs; find all blocked destination

IPs; and summarise events as a dict mapping action to count. Demonstrate accessing fields by name and by index, and show that instances are hashable by using one as a dict key.

2. Implement a role-based permission system using sets. Define at least five roles and their permission sets (use frozenset). Write functions: effective_permissions(roles) returns the union of all permissions; has_permission(roles, perm) checks if a permission is in the effective set; grant(user_perms, new_perms) returns a new set with additions; revoke(user_perms, removed_perms) returns a new set with removals; and compare_roles(role_a, role_b) prints what each role has that the other does not, and what they share.

3. Write a service change detector. Given two snapshots of a service registry (each a set of service names), compute: services added, services removed, services stable. Then extend it to detect configuration changes: given two snapshots where each service maps to a dict of config values, detect which services changed, what keys changed within each service, and what the old and new values were. Use set operations for all comparisons.

4. Build an O(1) IP allowlist/blocklist checker. Start with a blocklist of 50,000 IP addresses as a list. Measure the time to check 10,000 IPs against the list using in. Then convert to a set and repeat. Report the speedup factor. Extend the exercise to implement a three-tier lookup: if an IP is on the blocklist (set) return "blocked"; if on the allowlist (set) return "allowed"; otherwise return "unknown". Time the three-tier lookup on 10,000 IPs.

5. Write a tag-based resource filter system. Each resource in a catalogue has a set of tags. Implement a filter function that accepts: require_all (resource must have every listed tag), require_any (resource must have at least one), and exclude (resource must have none). Test with at least 10 resources and at least 5 different filter combinations that produce different result sets. Also implement a function that finds the set of tags appearing on every resource in a given result set (intersection of all tag sets).

Challenge Tasks

The following tasks are open-ended. No hints are provided. They require you to design a solution, not just apply a pattern.

1. Design and implement a dependency resolver using sets. You have a dict mapping package names to sets of their dependencies. Implement: resolve(package) returns the complete set of all transitive dependencies (direct and indirect); install_order(packages) returns a list of packages in an order where every package appears after all its dependencies (topological sort using sets); and detect_cycles(packages) finds any package that directly or transitively depends on itself. Demonstrate with a realistic package graph of at least 12 packages with at least 3 levels of transitive dependencies.

2. Build a session and token manager using sets and frozensets. User sessions are namedtuples (user_id, session_id, created_at, expires_at, scopes) where scopes is a frozenset of permissions. Implement: create_session, expire_session, get_active_sessions(user_id), revoke_scope(session_id, scope), check_access(session_id, required_scopes) (uses issubset), get_users_with_scope(scope) (across all active sessions), and find_expired_sessions() (compares current time to expires_at). Demonstrate all operations, including edge cases like checking a revoked scope and accessing an expired session.

3. Implement a data reconciliation tool. Given two datasets (each a list of records with a unique id field and several data fields), the tool should: identify records present in A but not B (using id sets); identify records in B but not A; identify records in both but with different field values (detect which fields changed); produce a merge result that takes the B version for updated records and the A version for records only in A; and report a full reconciliation summary with counts and change details. Handle at least 50 records per dataset with at least 5 deliberate differences to detect.

CHAPTER 14 Functions: Basics

Think, Code, Create – A Python Programming Journey

> A function is the fundamental unit of reuse, abstraction, and organisation in Python. It gives a name to a block of logic, separates what a computation does from how it does it, and allows the same behaviour to be invoked from multiple places with different inputs. Every non-trivial Python program is a collection of cooperating functions, and the quality of those functions — their names, their signatures, their documentation, and their scope discipline — directly determines the quality of the program.
>
> This chapter covers the complete basics: defining functions with type hints and docstrings, the full argument system (positional, keyword, default, *args, **kwargs), variable scope and the LEGB rule, recursion with proper memoisation using functools.lru_cache, lambda functions and when to use them, and function decorators including the functools.wraps pattern. All examples use professional domains rather than toy arithmetic.

14.1 Defining and Calling Functions

A function is defined with the def keyword, a name, a parameter list in parentheses, and a colon. The body is indented. The return statement sends a value back to the caller; a function without a return statement implicitly returns None. The function is called by writing its name followed by parentheses containing any arguments.

Two additions make functions significantly more useful and maintainable in professional code: type hints annotate each parameter and the return value with their expected types (not enforced at runtime, but checked by static tools and read by IDEs), and a docstring immediately inside the function body documents what the function does, its parameters, and its return value.

def, return, Type Hints, and Docstrings

Example 1 — Introduction

A well-structured function: type hints on all parameters and return, complete docstring, clear single responsibility.

```
def calculate_shipping_cost(
    weight_kg:    float,
    distance_km:  float,
    express:      bool = False,
) -> float:
    """
    Calculate the shipping cost for a parcel.

    Args:
        weight_kg:    Parcel weight in kilograms (must be > 0).
        distance_km: Shipping distance in kilometres (must be > 0).
        express:     If True, apply the 1.8x express multiplier.
```

```
    Returns:
        Shipping cost in AUD, rounded to two decimal places.

    Raises:
        ValueError: If weight_kg or distance_km is not positive.
    """
    if weight_kg <= 0 or distance_km <= 0:
        raise ValueError("weight_kg and distance_km must be positive")

    BASE_RATE   = 0.08   # dollars per kg per km
    EXPRESS_MUL = 1.8

    base = weight_kg * distance_km * BASE_RATE
    cost = base * EXPRESS_MUL if express else base
    return round(cost, 2)

print(calculate_shipping_cost(5.2, 312))            # 129.79
print(calculate_shipping_cost(5.2, 312, express=True))  # 233.63
```

Functions Without Return Values

A function that performs an action but does not produce a result for the caller to use should not return anything, or explicitly return None. Such functions are often called procedures. In type hints, the return type is annotated as -> None. Assigning the result of a None-returning function to a variable is a common mistake — it produces None, not the expected value.

Example 2 — Variation

A logging function that returns None, and the common mistake of assigning its result.

```
import sys
from datetime import datetime

def log_event(level: str, service: str, message: str) -> None:
    """Write a formatted log entry to stderr."""
    timestamp = datetime.now().strftime("%Y-%m-%dT%H:%M:%S")
    entry = f"[{timestamp}] {level.upper():<8} {service}: {message}"
    print(entry, file=sys.stderr)

# Correct usage: call for the side effect
log_event("info", "api-gateway", "Request received")
log_event("error", "auth",        "Invalid token")

# Common mistake: assigning the result of a None-returning function
result = log_event("warn", "cache", "Cache miss")
print(f"Result is: {result}")   # None -- not the log entry
```

Writing Good Docstrings

A docstring is a string literal immediately following the def line. It is accessible via help(function_name) and function_name.__doc__. The Google-style docstring format (shown in the example above) is widely used: a one-line summary, a blank line, then labelled sections for

Args, Returns, and Raises. Write docstrings for every function that is not entirely self-evident from its name and type hints.

Example 3 — Real-world application

Accessing docstrings programmatically and the benefits of help() in interactive development.

```
def parse_connection_string(conn_str: str) -> dict[str, str]:",
    """
    Parse a key=value connection string into a dictionary.

    Args:
        conn_str: A string of the form "host=db port=5432 dbname=inv".
            Pairs are separated by whitespace; keys and values by =.

    Returns:
        A dict mapping each key to its value (both as strings).

    Example:
        >>> parse_connection_string("host=db port=5432")
        {'host': 'db', 'port': '5432'}
    """
    result = {}
    for token in conn_str.split():
        if "=" in token:
            key, _, val = token.partition("=")
            result[key.strip()] = val.strip()
    return result

# Docstring is accessible at runtime
print(parse_connection_string.__doc__[:60])

cfg = parse_connection_string("host=db-prod port=5432 dbname=inv pool=10")
print(cfg)
```

14.2 Function Arguments

Python's argument system is more flexible than most languages. There are five distinct kinds of arguments — positional, keyword, positional-only (Python 3.8+), keyword-only, and variadic — and they can be combined in a single function signature. Understanding each kind and the rules for combining them enables you to design function interfaces that are both flexible and unambiguous.

Positional, Keyword, and Default Arguments

A positional argument is assigned by position: the first argument maps to the first parameter. A keyword argument is assigned by name: fn(param=value). Once a keyword argument appears in a call, all following arguments must also be keyword arguments. A default argument provides a fallback value used when the caller omits that argument; defaults must come after all non-default parameters in the signature.

Example 1 — Introduction

A report generation function demonstrating positional, keyword, and default arguments.

```
from datetime import date
```

```python
def generate_report(
    service:   str,
    metric:    str,
    start_date: date,
    end_date:  date  = None,
    format:    str   = "json",
    include_nulls: bool = False,
) -> dict:
    """Generate a metrics report for a service."""
    end = end_date or date.today()
    return {
        "service": service, "metric": metric,
        "start":   str(start_date), "end": str(end),
        "format":  format, "include_nulls": include_nulls,
    }

# Positional: first three required args in order
r1 = generate_report("api", "latency", date(2024, 7, 1))

# Mix positional and keyword: optional args by name
r2 = generate_report("api", "latency", date(2024, 7, 1),
                     format="csv", include_nulls=True)

# All keyword: most readable for many parameters
r3 = generate_report(
    service="db", metric="connections",
    start_date=date(2024, 7, 1), end_date=date(2024, 7, 31),
)
print(r3)
```

*args and **kwargs

*args collects any number of extra positional arguments into a tuple named args. **kwargs collects any number of extra keyword arguments into a dict named kwargs. These are used when the function must accept a variable number of inputs whose count is not known at definition time. The conventional names args and kwargs are not required but are expected by Python developers. The ordering rule is: positional params, *args, keyword-only params, **kwargs.

Important: do not use a mutable object (list, dict) as a default parameter value. The default is created once when the function is defined, not each time it is called. Mutations to the default persist between calls, producing surprising results. Use None as the default and create the mutable object inside the function body.

Example 2 — Variation

**args for a variadic cost accumulator; **kwargs for a flexible config builder; the mutable default anti-pattern.*

```python
# *args: accept any number of positional line-item amounts
def calculate_invoice_total(*amounts: float, tax_rate: float = 0.10) -> float:
    """Sum all amounts and apply tax. tax_rate is keyword-only after *."""
    subtotal = sum(amounts)
    return round(subtotal * (1 + tax_rate), 2)
```

```python
print(calculate_invoice_total(120.00, 34.95, 8.50))
print(calculate_invoice_total(120.00, 34.95, tax_rate=0.0))

# **kwargs: flexible config dict builder
def build_config(**overrides: object) -> dict:
    """Merge overrides into a base configuration."""
    base = {"timeout": 30, "retries": 3, "log_level": "INFO", "debug": False}
    return {**base, **overrides}

cfg = build_config(timeout=60, debug=True, host="db-prod")
print(cfg)

# ANTI-PATTERN: mutable default argument
def add_tag_bad(tag, tags=[]):    # Default [] is shared across all calls!
    tags.append(tag)
    return tags

# CORRECT: use None and create inside the function
def add_tag_good(tag, tags=None):
    if tags is None:
        tags = []
    tags.append(tag)
    return tags
```

14.3 Scope and Variable Lifetime

Scope determines which names are visible from which parts of a program. Python resolves names using the LEGB rule: Local → Enclosing → Global → Built-in. Python searches each scope in that order and uses the first match found. Understanding this rule explains why code behaves as it does when names appear at different levels.

The LEGB Rule

Local scope: names defined inside the current function. Enclosing scope: names defined in any enclosing function (for nested functions). Global scope: names defined at module level. Built-in scope: Python's built-in names (len, print, range, etc.). A name in an inner scope shadows the same name in an outer scope without modifying the outer one.

Example 1 — Introduction

The LEGB lookup chain demonstrated with a configuration resolver.

```python
LOG_LEVEL = "INFO"        # Global scope

def configure_service(service_name: str):
    LOG_LEVEL = "DEBUG"   # Enclosing scope (shadows global)

    def get_log_config() -> dict:
        timeout = 30      # Local scope
        # LEGB: timeout (L) -> LOG_LEVEL (E) -> ... -> len (B)
        return {
            "service":   service_name,    # E: enclosing
            "log_level": LOG_LEVEL,       # E: enclosing (not global!)
            "timeout":   timeout,         # L: local
```

```
        }

    config = get_log_config()
    print(f"  Service config: {config}")

configure_service("auth")
print(f"Global LOG_LEVEL unchanged: {LOG_LEVEL}")
```

global and nonlocal

The global keyword declares that a name inside a function refers to the module-level variable, allowing it to be modified. Using global is strongly discouraged in production code — it creates hidden dependencies between functions and makes programs harder to test and reason about. The same goal is almost always achievable by passing the value as a parameter and returning the result.

nonlocal declares that a name in an inner function refers to the variable in the nearest enclosing (non-global) scope. This is legitimate and useful in closures — functions that maintain state via a captured variable from their enclosing scope.

Example 2 — Variation

global (discouraged) vs parameter-passing (preferred); nonlocal used legitimately in a counter closure.

```
# DISCOURAGED: global variable modified in-place
total_errors = 0
def record_error_bad():
    global total_errors    # hidden dependency on global state
    total_errors += 1

# PREFERRED: pass state in, return new state
def record_error_good(current_count: int) -> int:
    return current_count + 1

count = 0
count = record_error_good(count)
count = record_error_good(count)
print(f"Error count: {count}")

# nonlocal is legitimate: counter closure
def make_counter(start: int = 0):
    """Return a function that increments and returns a private counter."""
    count = start

    def increment(by: int = 1) -> int:
        nonlocal count
        count += by
        return count

    return increment

request_counter = make_counter()
print(request_counter())    # 1
print(request_counter())    # 2
print(request_counter(10))  # 12
```

14.4 Recursion

A recursive function calls itself as part of its own definition. Every recursive function needs a base case (a condition that stops the recursion and returns a concrete value) and a recursive case (a call to itself with a simpler or smaller version of the problem). Without a base case, the function calls itself indefinitely until Python raises a RecursionError (the default stack depth limit is 1000).

Recursion is natural for problems with self-similar structure: tree traversal, nested data processing, divide-and-conquer algorithms. For simple counting loops, an iterative approach is usually clearer and faster.

Base Case and Recursive Case

Example 1 — Introduction

Recursive directory size calculator: natural recursion over a tree structure.

```
import os

def directory_size(path: str) -> int:
    """
    Recursively compute the total size (bytes) of a directory tree.

    Args:
        path: Path to a file or directory.

    Returns:
        Total size in bytes.
    """
    # Base case: a file -- return its size
    if os.path.isfile(path):
        return os.path.getsize(path)

    # Recursive case: a directory -- sum all children recursively
    total = 0
    try:
        for entry in os.scandir(path):
            total += directory_size(entry.path)   # recursive call
    except PermissionError:
        pass
    return total

size = directory_size(".")
print(f"Current directory: {size:,} bytes ({size/1024:.1f} KB)")
```

Memoisation with functools.lru_cache

Some recursive functions recompute the same sub-problems repeatedly (overlapping subproblems). Memoisation caches results so each unique input is computed only once. The correct modern approach is functools.lru_cache, not a mutable default argument (def f(n, memo={}) is an anti-pattern: the dict is shared across all calls and can cause subtle bugs). lru_cache is applied as a decorator and handles caching transparently.

Example 2 — Variation

functools.lru_cache for memoised recursion vs the anti-pattern mutable default.

```
from functools import lru_cache

# CORRECT: lru_cache decorator handles memoisation cleanly
@lru_cache(maxsize=None)
def fibonacci(n: int) -> int:
    """Return the nth Fibonacci number (0-indexed, F(0)=0, F(1)=1)."""
    if n < 0:
        raise ValueError(f"n must be non-negative, got {n}")
    if n <= 1:
        return n
    return fibonacci(n - 1) + fibonacci(n - 2)

print([fibonacci(i) for i in range(10)])
# [0, 1, 1, 2, 3, 5, 8, 13, 21, 34]

# Cache stats: see how many calls were actually made
print(fibonacci.cache_info())

# ANTI-PATTERN (do NOT do this):
# def fibonacci_bad(n, memo={}):   # mutable default shared across calls
#     ...                           # can cause hard-to-debug state issues
```

Recursive Data Processing

Example 3 — Real-world application

Recursive flattening of an arbitrarily nested configuration structure.

```
def flatten_nested(data: object, prefix: str = "") -> dict:
    """
    Recursively flatten a nested dict/list structure.
    Lists become indexed: parent.0, parent.1, ...
    Dicts become dotted: parent.key.
    """
    result = {}
    if isinstance(data, dict):
        for key, value in data.items():
            new_key = f"{prefix}.{key}" if prefix else key
            result.update(flatten_nested(value, new_key))
    elif isinstance(data, list):
        for i, item in enumerate(data):
            new_key = f"{prefix}.{i}" if prefix else str(i)
            result.update(flatten_nested(item, new_key))
    else:
        result[prefix] = data    # base case: scalar value
    return result


config = {
    "database": {"host": "db-01", "port": 5432},
    "services": ["api", "auth", "cache"],
    "debug": False,
}
for key, val in flatten_nested(config).items():
    print(f"  {key:<30} = {val}")
```

14.5 Lambda Functions and Functional Tools

A lambda is an anonymous inline function: lambda parameters: expression. It is a single expression — no statements, no return keyword, no docstring. Lambdas are most useful as the key= argument to sorted(), min(), max(), and groupby(), and as short callback functions. For anything more than a single expression, define a named function with def — it is clearer, testable, and debuggable.

Python has map() and filter() but list comprehensions and generator expressions are almost always more readable for the same task in Python 3. Use map() and filter() when you have an already-named function to apply; use a comprehension when writing the logic inline.

Lambda as Sort Key

Example 1 — Introduction

Lambda as key= for sorting, min(), and max() on structured records.

```
incidents = [
    {"id": "INC-003", "severity": 2, "service": "auth",  "age_min": 45},
    {"id": "INC-001", "severity": 1, "service": "api",   "age_min": 120},
    {"id": "INC-004", "severity": 1, "service": "auth",  "age_min": 10},
    {"id": "INC-002", "severity": 3, "service": "cache", "age_min": 5},
]

# Sort by severity descending, then age descending (oldest first)
ranked = sorted(incidents,
                key=lambda i: (-i["severity"], -i["age_min"]))

print("Incident queue:")
for inc in ranked:
    print(f"  [{inc['severity']}] {inc['id']} {inc['service']}
({inc['age_min']}min)")

# min/max with lambda key
oldest  = max(incidents, key=lambda i: i["age_min"])
highest = max(incidents, key=lambda i: i["severity"])
print(f"Oldest:   {oldest['id']} ({oldest['age_min']} min)")
print(f"Highest:  {highest['id']} (severity {highest['severity']})")
```

functools.partial for Partial Application

functools.partial creates a new callable by fixing some arguments of an existing function. This is useful for creating specialised versions of general functions without writing a new wrapper function, and for adapting function signatures to interfaces that require a specific number of arguments.

Example 2 — Variation

functools.partial creating specialised log functions from a general one.

```
from functools import partial
import sys
```

```python
def log(level: str, service: str, message: str, file=sys.stderr) -> None:
    """General-purpose structured log function."""
    print(f"[{level.upper():<8}] {service}: {message}", file=file)

# Create specialised versions by fixing the level argument
log_error = partial(log, "error")
log_warn  = partial(log, "warn")
log_info  = partial(log, "info")

# Specialised functions have the same interface but fixed level
log_error("auth-service", "Token validation failed")
log_warn("cache",         "Memory threshold at 87%")
log_info("api-gateway",   "Service started on port 8443")

# partial also works for adapting signatures
from functools import reduce
multiply = lambda x, y: x * y
product  = partial(reduce, multiply)
print(product([2, 3, 4, 5]))  # 120
```

map(), filter(), and When to Prefer Comprehensions

Example 3 — Real-world application

map() and filter() with named functions vs equivalent comprehensions — when each is clearer.

```python
records = [
    {"sku": "P-001", "price": "  12.50 ", "stock": "45", "active": "true"},
    {"sku": "P-002", "price": "  89.99 ", "stock": "0",  "active": "true"},
    {"sku": "P-003", "price": "  34.95 ", "stock": "12", "active": "false"},
]

# Named function: map() is readable when the function already exists
def parse_record(r: dict) -> dict:
    return {
        "sku":    r["sku"],
        "price":  float(r["price"].strip()),
        "stock":  int(r["stock"]),
        "active": r["active"].lower() == "true",
    }

parsed = list(map(parse_record, records))  # clear: function + map()

# Filter with named predicate: also clear
def is_available(r: dict) -> bool:
    return r["active"] and r["stock"] > 0

available = list(filter(is_available, parsed))

# Inline logic: comprehension is cleaner than lambda + map/filter
prices    = [r["price"] for r in parsed]              # vs map(lambda...)
in_stock  = [r for r in parsed if r["stock"] > 0]    # vs filter(lambda...)
print(f"Available: {len(available)}, Prices: {prices}")
```

14.6 Type Hints and Docstrings

Type hints annotate function signatures with the expected types of parameters and return values. They are not enforced at runtime, but are checked by static analysis tools (mypy, pyright), surfaced in IDE autocompletion, and serve as machine-readable documentation. The typing module provides constructs for composite types: Optional[T] for a value that may be None, Union[A, B] for multiple acceptable types, and in Python 3.10+ the | syntax for both.

Annotating Parameters and Return Types

Example 1 — Introduction

A fully annotated function using standard types and typing module constructs.

```
from typing import Optional

# Python 3.9+: built-in generic types (list, dict, tuple)
# Python 3.10+: X | Y union syntax instead of Union[X, Y]

def process_transactions(
    transactions: list[dict],
    region_filter: Optional[str] = None,
    min_amount:    float         = 0.0,
    max_results:   int           = 100,
) -> tuple[list[dict], dict[str, float]]:
    """
    Filter and summarise transactions.

    Args:
        transactions:  List of transaction dicts with amount, region, status.
        region_filter: If provided, only include this region.
        min_amount:    Minimum transaction amount to include.
        max_results:   Maximum number of results to return.

    Returns:
        A tuple of (filtered_records, summary) where summary contains
        total, mean, and count as floats.
    """
    filtered = [
        t for t in transactions
        if t["amount"] >= min_amount
        and (region_filter is None or t.get("region") == region_filter)
    ][:max_results]

    amounts = [t["amount"] for t in filtered]
    summary = {
        "total": sum(amounts),
        "mean":  sum(amounts) / len(amounts) if amounts else 0.0,
        "count": float(len(amounts)),
    }
    return filtered, summary
```

Optional, Union, and Generic Types

Optional[T] is shorthand for Union[T, None] and means "this parameter/return value might be None". In Python 3.10+ you can write T | None instead. list[str] annotates a list of strings;

dict[str, int] annotates a dict mapping str to int; tuple[int, str, float] annotates a fixed-length tuple with those exact types. These generic forms (using [] not [()]) are available on the built-in types from Python 3.9.

Example 2 — Variation

Optional, Union, and generic container types in a configuration loader.

```
from typing import Union

ConfigValue = Union[str, int, float, bool, None]

def get_config_value(
    config: dict[str, ConfigValue],
    key:    str,
    default: ConfigValue = None,
    cast_to: type | None = None,
) -> ConfigValue:
    """
    Retrieve a configuration value, optionally casting it.

    Args:
        config:   Configuration dictionary.
        key:      The key to retrieve.
        default:  Value returned if key is absent.
        cast_to:  If provided, cast the value to this type before returning.

    Returns:
        The configuration value or default, optionally cast.
    """
    value = config.get(key, default)
    if cast_to is not None and value is not None:
        try:
            return cast_to(value)
        except (ValueError, TypeError):
            return default
    return value

cfg = {"timeout": "30", "debug": "true", "max_conn": "20"}
print(get_config_value(cfg, "timeout", default=10, cast_to=int))   # 30
print(get_config_value(cfg, "missing", default=0,  cast_to=float)) # 0
```

14.7 Function Decorators

A decorator is a function that takes a function as its argument and returns a new function that wraps the original, adding behaviour before or after the call. Applied with @decorator_name above the function definition, it is equivalent to fn = decorator_name(fn). Decorators are the standard mechanism for cross-cutting concerns: logging, timing, retry logic, access control, and caching.

Writing a Basic Decorator

Every decorator follows the same structure: an outer function accepts the decorated function, an inner wrapper function adds behaviour and calls the original, and the outer function returns the wrapper. functools.wraps must be applied to the wrapper — without it, the decorated function loses its name, docstring, and other metadata, which breaks introspection and help().

Example 1 — Introduction

A timing decorator using functools.wraps to preserve function metadata.

```
import time
from functools import wraps

def timed(func):
    """Decorator: log the execution time of the decorated function."""
    @wraps(func)          # preserves __name__, __doc__, etc.
    def wrapper(*args, **kwargs):
        start  = time.perf_counter()
        result = func(*args, **kwargs)
        elapsed = time.perf_counter() - start
        print(f"  {func.__name__}() completed in {elapsed*1000:.2f}ms")
        return result
    return wrapper

@timed
def load_records(source: str, limit: int = 1000) -> list:
    """Simulate loading records from a data source."""
    time.sleep(0.05)   # simulate I/O
    return [{"id": i} for i in range(limit)]

records = load_records("transactions", limit=500)
print(f"  Loaded {len(records)} records")

# @wraps preserves the original metadata
print(f"  Function name: {load_records.__name__}")  # load_records, not
wrapper
```

Decorators with Arguments

A decorator that needs its own arguments requires one extra layer of wrapping: a function that accepts the arguments and returns the actual decorator. The @ syntax then calls the outer function, which returns the decorator, which is applied to the function.

Example 2 — Variation

A retry decorator that accepts max_attempts and delay_seconds as arguments.

```
import time
from functools import wraps

def retry(max_attempts: int = 3, delay_s: float = 1.0,
exceptions=(Exception,)):
    """
    Decorator factory: retry the decorated function on exception.

    Args:
        max_attempts: Maximum number of attempts before raising.
        delay_s:      Seconds to wait between attempts.
        exceptions:   Tuple of exception types to catch and retry.
    """
    def decorator(func):
```

```
        @wraps(func)
        def wrapper(*args, **kwargs):
            last_exc = None
            for attempt in range(1, max_attempts + 1):
                try:
                    return func(*args, **kwargs)
                except exceptions as e:
                    last_exc = e
                    if attempt < max_attempts:
                        print(f"  {func.__name__}: attempt {attempt} failed: 
{e}. Retrying...")
                        time.sleep(delay_s)
            raise last_exc
        return wrapper
    return decorator

import random

@retry(max_attempts=4, delay_s=0.1, exceptions=(ConnectionError,))
def call_api(endpoint: str) -> dict:
    """Simulate an unreliable API call."""
    if random.random() < 0.6:
        raise ConnectionError(f"Timeout calling {endpoint}")
    return {"status": "ok", "endpoint": endpoint}

try:
    result = call_api("/api/inventory")
    print(f"  Success: {result}")
except ConnectionError as e:
    print(f"  All attempts failed: {e}")
```

Chapter 14 Summary

- Define functions with def, a name, typed parameters, and a return type annotation. Type hints are not enforced at runtime but are checked by static tools and IDE autocompletion. A docstring immediately inside the function body documents purpose, args, returns, and raises. Functions without a meaningful return value should have return type -> None.
- Python has five argument kinds: positional (assigned by position), keyword (assigned by name), default (used when caller omits the argument), *args (collects extra positional args into a tuple), and **kwargs (collects extra keyword args into a dict). Defaults must come after required params. NEVER use a mutable object as a default value — use None and create the object inside the function.
- Python resolves names using the LEGB rule: Local → Enclosing → Global → Built-in. Avoid modifying global state with the global keyword; pass values in as parameters and return them instead. nonlocal is legitimate in closures where an inner function must update a captured variable from its enclosing scope.
- Recursive functions need a base case (stops recursion) and a recursive case (calls itself with a simpler input). Use functools.lru_cache to memoise overlapping subproblems — never use mutable default arguments (def f(n, memo={}) is an anti-pattern). Python's default stack limit is 1000 frames; very deep recursion raises RecursionError.

- Lambda functions (lambda params: expression) are anonymous, single-expression callables. Use them as short key= arguments to sorted(), min(), max(). Use functools.partial to create specialised functions by fixing some arguments of a general function. Prefer list comprehensions over lambda + map()/filter() for inline logic; use map() with a named function when the function already exists.
- Type hints: Optional[T] (or T | None in Python 3.10+) for possibly-None values; Union[A, B] for multiple types; list[T], dict[K, V], tuple[T, ...] for generic containers. The typing module provides these constructs; in Python 3.9+ the built-in types support subscripting directly.
- Decorators wrap a function to add behaviour. Structure: outer function accepts the decorated function, inner wrapper function adds behaviour and calls original, outer returns wrapper. Always apply @functools.wraps to the wrapper to preserve the original function's metadata. Decorator factories (for decorators with arguments) add one extra outer function.

Guided Exercises

Complete the following exercises in VS Code, IDLE, or your preferred Python environment. Each exercise targets a specific concept from this chapter.

1. Write a fully typed and documented function library for financial calculations. Include at least four functions: calculate_compound_interest(principal, rate, compounds_per_year, years) -> float; calculate_loan_payment(principal, annual_rate, term_years) -> float; apply_tax(amount, rate, round_up=False) -> float; and format_currency(amount, currency="AUD", decimals=2) -> str. Each function must have complete type hints, a Google-style docstring with Args/Returns/Raises, and input validation that raises ValueError with a descriptive message for invalid inputs.

2. Demonstrate the mutable default argument anti-pattern and its fix. Write two versions of a function that accumulates items into a list: one using a mutable default, one using None. Call each version three times in a row with a single argument and show how the results differ. Then write a third version using *args to accept any number of items in a single call, and a fourth that accepts an optional items_list parameter. Test all four versions.

3. Write a closure-based counter factory that uses nonlocal. The factory should accept a start value, a step, and an optional maximum. The returned function should increment by step on each call, raise StopIteration when the maximum is reached (if set), and support a reset() method to return to start. Demonstrate with three different counter instances running independently of each other, confirming that each has its own private state.

4. Build a recursive JSON-like data summariser. Write a function analyse(data) that recursively inspects any combination of dicts, lists, and scalar values and returns a summary dict containing: the total number of nodes (dicts + lists + scalars counted separately), the maximum nesting depth, all unique scalar types found (as a set of type names), and all unique dict keys found at any depth. Test with a nested structure at least four levels deep containing all supported types.

5. Create a timing and logging decorator system with two decorators: @timed logs the function name and execution time; @logged logs the arguments and return value. Both must use functools.wraps. Write a third decorator @retry(attempts, delay) that retries on exception. Demonstrate all three applied to the same function (using stacked decorators), and show the order of execution. Also show that functools.wraps preserves __name__ and __doc__ correctly.

Challenge Tasks

The following tasks are open-ended. No hints are provided. They require you to design a solution, not just apply a pattern.

1. Design and implement a type-safe configuration system using typed functions. Define a ConfigSchema as a dict mapping key names to descriptors (type, required, default, validator callable). Write load_config(schema, raw_dict) that: validates all required keys are present, casts each value to the declared type, runs any validator, collects ALL errors before raising, and returns a clean typed dict. Write a typed accessor get_typed(config, key, type_) that returns the value or raises TypeError if the type does not match. Demonstrate with at least three schemas of different complexity.

2. Build a function pipeline using functools.partial and lambda. Implement a Pipeline class whose constructor accepts a list of callables. Calling the pipeline on an input passes the result of each function to the next. Support: adding functions with pipe(fn), conditional application with pipe_if(condition, fn) (skips fn if condition is False), and a fork(fn_a, fn_b) operation that returns a tuple of results from two separate pipelines. Demonstrate with a realistic data-cleaning pipeline that normalises, validates, transforms, and formats records.

3. Implement a memoisation decorator more powerful than functools.lru_cache. Your @memoize decorator should: support a configurable TTL (time-to-live) in seconds after which cached entries expire; support both positional and keyword arguments (using a canonical cache key); track hit rate (hits vs total calls); expose a cache_info() method returning hits, misses, size, and hit_rate; and provide an invalidate(*args, **kwargs) method to remove a specific entry. Demonstrate with at least two functions: one pure (factorial), one time-dependent (simulated API call with random latency). Verify TTL expiry, show hit rate stats, and test cache invalidation.

CHAPTER 15 Functions: Advanced Use

Think, Code, Create – A Python Programming Journey

> Chapter 14 covered the mechanics of defining and calling functions. This chapter covers what becomes possible once you treat functions as values — objects that can be stored, passed, returned, and combined like any other data. Python's first-class functions are the foundation of decorators, closures, generators, and functional data-processing pipelines. Together, these techniques enable a style of programming that is concise, composable, and highly testable.
>
> This chapter covers: functions as first-class objects; closures and factory functions for encapsulating configurable behaviour; generator functions for lazy evaluation of large or infinite sequences; function composition and processing pipelines; principles for structuring larger programs with functions; and a brief introduction to methods as functions that live inside classes.

15.1 Functions as First-Class Objects

In Python, functions are first-class objects: they can be assigned to variables, stored in data structures, passed as arguments to other functions, and returned as values from functions. A function that accepts or returns another function is called a higher-order function. This is not a special feature — it follows directly from Python's object model where everything, including a function, is an object.

Storing Functions in Collections

Because functions are objects, they can be placed in lists, dicts, or any other container. A dict of functions is the dispatch table pattern from Chapter 12 applied at full power: the table selects the handler, the handler performs the work, and adding a new case requires only a new dict entry.

Example 1 — Introduction

Functions stored in a dict for a configurable data transformation pipeline.

```
from decimal import Decimal, ROUND_HALF_UP

# Functions are just objects -- store them in a dict
TRANSFORMERS = {
    "strip":      str.strip,
    "lower":      str.casefold,
    "upper":      str.upper,
    "title":      str.title,
    "to_int":     int,
    "to_float":   float,
    "to_decimal": lambda v: Decimal(v).quantize(Decimal("0.01"),
rounding=ROUND_HALF_UP),
}
```

```
def transform(value: str, *operations: str) -> object:
    """Apply a sequence of named transformations to a value."""
    result = value
    for op in operations:
        fn = TRANSFORMERS.get(op)
        if fn is None:
            raise ValueError(f"Unknown operation: {op!r}")
        result = fn(result)
    return result

print(transform("  HELLO WORLD  ", "strip", "lower"))  # "hello world"
print(transform("  49.955  ",      "strip", "to_decimal")) # Decimal('49.96')
print(transform("  Api Gateway  ", "strip", "lower"))      # "api gateway"
```

Functions as Arguments (Callbacks)

Passing a function as an argument allows the caller to customise behaviour without modifying the function receiving it. This is the callback pattern: the higher-order function defines the structure of the operation; the caller supplies the logic. sorted(), max(), min(), filter(), and map() are all built-in higher-order functions that accept a function argument.

Example 2 — Variation

A reusable retry function that accepts the operation to retry as an argument.

```
import time, random
from typing import Callable, TypeVar

T = TypeVar("T")

def with_retry(
    operation:    Callable[[], T],
    max_attempts: int   = 3,
    delay_s:      float = 1.0,
    backoff:      float = 2.0,
) -> T:
    """
    Execute operation(), retrying on any exception.
    Delay doubles on each failure (exponential backoff).
    """
    delay = delay_s
    for attempt in range(1, max_attempts + 1):
        try:
            return operation()
        except Exception as e:
            if attempt == max_attempts:
                raise
            print(f"  Attempt {attempt} failed: {e}. Retrying in
{delay:.1f}s...")
            time.sleep(delay)
            delay *= backoff

def flaky_api_call() -> dict:
    if random.random() < 0.6:
        raise ConnectionError("Timeout")
    return {"status": "ok"}
```

```
result = with_retry(flaky_api_call, max_attempts=4, delay_s=0.1)
print(f"Result: {result}")
```

Functions as Return Values

Example 3 — Real-world application

A function that returns a configured validator function.

```
from typing import Callable

def make_range_validator(
    min_val: float,
    max_val: float,
    field_name: str = "value",
) -> Callable[[float], None]:
    """
    Return a validator function that raises ValueError if the
    input is outside [min_val, max_val].
    """
    def validate(value: float) -> None:
        if not (min_val <= value <= max_val):
            raise ValueError(
                f"{field_name} must be between {min_val} and {max_val}, got
{value}"
            )
    return validate

# Create specialised validators
validate_port     = make_range_validator(1, 65_535, "port")
validate_cpu_pct  = make_range_validator(0, 100,    "cpu_pct")
validate_timeout  = make_range_validator(1, 300,    "timeout_s")

validate_port(8443)      # OK
validate_cpu_pct(87.5)   # OK
try:
    validate_port(99_999)
except ValueError as e:
    print(f"Error: {e}")
```

13.2 Closures and Factory Functions

A closure is a function that captures variables from its enclosing scope. When the outer function returns the inner function, the inner function continues to hold a reference to those captured variables — even after the outer function has finished executing. This gives the inner function a private, persistent state without requiring a class.

A factory function is an outer function that creates and returns a configured inner function. Each call to the factory produces a distinct closure with its own captured variables. This is the clean alternative to global state: instead of a shared mutable global, each instance of the closure has its own private data.

Rate Limiter Closure

Example 1 — Introduction

A rate limiter factory: each returned function enforces an independent request limit.

```
import time

def make_rate_limiter(calls_per_second: float) -> callable:
    """
    Return a function that blocks until the rate limit allows the next call.
    Each call to make_rate_limiter() creates an independent rate limiter.
    """
    min_interval = 1.0 / calls_per_second
    last_call_time = [0.0]    # list to allow mutation inside inner function

    def rate_limited_call() -> None:
        now     = time.monotonic()
        elapsed = now - last_call_time[0]
        if elapsed < min_interval:
            time.sleep(min_interval - elapsed)
        last_call_time[0] = time.monotonic()

    return rate_limited_call

# Two independent limiters -- each with its own state
api_limiter     = make_rate_limiter(5)    # 5 calls per second
scraper_limiter = make_rate_limiter(0.5)  # 1 call every 2 seconds

for i in range(3):
    api_limiter()
    print(f"API call {i+1} at {time.monotonic():.2f}s")
```

Memoisation Closure

A manual memoisation closure stores computed results in a dict captured from the enclosing scope. Each call checks the cache first; if the result is present it is returned immediately. This demonstrates how closures can maintain private mutable state across calls, and why functools.lru_cache is the recommended approach in production (it handles thread safety and bounded cache sizes that a hand-rolled closure does not).

Example 2 — Variation

Hand-rolled memoisation closure showing the cache mechanism; lru_cache is preferred in practice.

```
def make_memoised(func):
    """
    Return a memoised version of func.
    The cache dict is captured from the closure scope.
    """
    cache = {}       # private to this closure instance
    call_count = [0]

    def memoised(*args):
        if args not in cache:
            call_count[0] += 1
            cache[args] = func(*args)
```

```
        return cache[args]

    memoised.cache      = cache       # expose for inspection
    memoised.call_count = call_count
    return memoised

def expensive_query(region: str, metric: str) -> float:
    """Simulate an expensive database query."""
    import time, random
    time.sleep(0.05)
    return round(random.uniform(10, 100), 2)

cached_query = make_memoised(expensive_query)

for region in ["AU", "NZ", "AU", "SG", "NZ", "AU"]:",
    result = cached_query(region, "latency")
    print(f"  {region}: {result}")

print(f"DB calls made: {cached_query.call_count[0]} (cached {6 -
cached_query.call_count[0]})")
```

Currying and Partial Application via Closures

Example 3 — Real-world application

Building a configurable formatter using currying: fix some arguments, produce specialised functions.

```
def make_formatter(template: str, currency: str = "AUD"):
    """Return a formatter function for a specific template and currency."""
    def format_value(value: float) -> str:
        return template.format(value=value, currency=currency)
    return format_value

# Create specialised formatters by fixing template and currency
fmt_price    = make_formatter("${value:,.2f} {currency}")
fmt_tax      = make_formatter("  Tax ({currency}): ${value:,.2f}")
fmt_percent  = make_formatter("{value:.1f}%")
fmt_usd      = make_formatter("${value:,.2f} {currency}", "USD")

items = [("Widget A", 149.95, 0.10), ("Gadget B", 89.99, 0.10)]

for name, price, rate in items:
    print(f"  {name}")
    print(f"    Price:  {fmt_price(price)}")
    print(f"    {fmt_tax(price * rate)}")
    print(f"    Rate:   {fmt_percent(rate * 100)}")
    print(f"    USD:    {fmt_usd(price * 0.65)}")
```

15.3 Generator Functions

A generator function contains a yield statement. When called, it does not execute immediately — it returns a generator object. Each call to next() on the generator runs the function body until the next yield, which produces a value and pauses execution. When the function returns or falls off the end, the generator raises StopIteration.

Generators produce values lazily: one at a time, on demand, without storing the full sequence in memory. This makes them ideal for large datasets, infinite sequences, and streaming data. A for loop calls next() automatically and handles StopIteration.

Basic Generator Functions

Example 1 — Introduction

A log-line generator that reads and filters a file one line at a time without loading it all.

```
from datetime import datetime

def error_log_reader(lines: list[str], min_level: str = "ERROR") -> object:
    """
    Generator: yield parsed records for lines at or above min_level.
    Processes lines lazily -- suitable for very large log files.
    """
    LEVELS = {"DEBUG": 0, "INFO": 1, "WARN": 2, "WARNING": 2,
              "ERROR": 3, "CRITICAL": 4}
    min_rank = LEVELS.get(min_level.upper(), 3)

    for line in lines:
        parts = line.strip().split(None, 3)
        if len(parts) < 3:
            continue
        level = parts[1].upper()
        if LEVELS.get(level, -1) >= min_rank:
            yield {"timestamp": parts[0], "level": level,
                   "service": parts[2] if len(parts) > 3 else "",
                   "message": parts[3] if len(parts) > 3 else parts[2]}

# Simulate a log file as a list of lines
log_lines = [
    "2024-07-15T09:14 INFO  api-gw  Request OK",
    "2024-07-15T09:14 ERROR auth    Token expired",
    "2024-07-15T09:15 WARN  cache   Memory 88%",
    "2024-07-15T09:15 CRITICAL db   Connection lost",
    "not-a-valid-line",
]

for record in error_log_reader(log_lines, min_level="WARN"):
    print(f"  [{record['level']}] {record['timestamp']}: {record['message']}")
```

Infinite Generators

Because generators produce values on demand rather than all at once, they can represent infinite sequences. Infinite generators are consumed safely by taking only as many values as needed using next(), itertools.islice(), or a for loop with a break condition.

Example 2 — Variation

An infinite Fibonacci generator and an exponential backoff delay generator.

```
def fibonacci_gen():
    """Infinite Fibonacci sequence generator."""
    a, b = 0, 1
    while True:
        yield a
```

```
        a, b = b, a + b

def exponential_backoff(
    initial_delay: float = 0.5,
    multiplier:    float = 2.0,
    max_delay:     float = 60.0,
):"
    """Yield increasing delays for retry logic."""
    delay = initial_delay
    while True:
        yield delay
        delay = min(delay * multiplier, max_delay)

# Take only the first 10 Fibonacci numbers
import itertools
first_10 = list(itertools.islice(fibonacci_gen(), 10))
print(f"Fibonacci: {first_10}")

# Use backoff delays in a retry loop
backoff = exponential_backoff(initial_delay=0.1, max_delay=5.0)
for attempt in range(1, 6):
    delay = next(backoff)
    print(f"  Attempt {attempt}: wait {delay:.1f}s before retry")
```

Generator Pipelines

Multiple generators can be chained into a pipeline: each generator consumes from the previous one and yields transformed values. Only one element moves through the entire pipeline at a time, making the memory cost O(1) regardless of the dataset size.

Example 3 — Real-world application

A three-stage generator pipeline: read → parse → filter, processing records one at a time.

```
def read_records(raw_data: list[dict]) -> object:
    """Stage 1: yield raw records one at a time (simulates streaming
source)."""
    for record in raw_data:
        yield record

def parse_amounts(records) -> object:
    """Stage 2: convert amount strings to floats, skip unparseable."""
    for r in records:
        try:
            yield {**r, "amount": float(r["amount"])}
        except (ValueError, KeyError):
            pass   # silently skip malformed records

def filter_by_region(records, regions: set) -> object:
    """Stage 3: yield only records whose region is in the approved set."""
    for r in records:
        if r.get("region") in regions:
            yield r

raw = [
    {"id": "T-001", "amount": "149.95", "region": "AU"},
    {"id": "T-002", "amount": "N/A",    "region": "NZ"},
```

```
    {"id": "T-003", "amount": "89.99",  "region": "SG"},
    {"id": "T-004", "amount": "320.00", "region": "AU"},
]

# Chain generators into a pipeline -- O(1) memory
APPROVED = {"AU", "NZ"}
pipeline = filter_by_region(parse_amounts(read_records(raw)), APPROVED)

total = 0
for record in pipeline:
    total += record["amount"]
    print(f"  {record['id']}: ${record['amount']:.2f}")
print(f"Total: ${total:.2f}")
```

15.4 Function Composition and Pipelines

Function composition combines multiple functions into one: the output of each function becomes the input of the next. It is the functional-programming equivalent of method chaining and is particularly useful for building data-processing pipelines from reusable, single-purpose functions. Each function in the pipeline does one thing; the composition does everything.

Manual Composition

Example 1 — Introduction

Manual function composition applied to a data normalisation pipeline.

```
from typing import Callable

def compose(*fns: Callable) -> Callable:
    """
    Compose functions left-to-right: compose(f, g, h)(x) == h(g(f(x))).
    """
    def composed(value):
        for fn in fns:
            value = fn(value)
        return value
    return composed

# Individual single-purpose transformations
def strip_whitespace(s: str) -> str:  return s.strip()
def to_lowercase(s: str)     -> str:  return s.casefold()
def remove_punctuation(s: str) -> str:
    return "".join(c for c in s if c.isalnum() or c.isspace())
def collapse_spaces(s: str)  -> str:  return " ".join(s.split())

# Compose into a single normaliser
normalise = compose(strip_whitespace, to_lowercase,
                    remove_punctuation, collapse_spaces)

inputs = [
    "  Acme Corp.  ",
    "GLOBAL WIDGETS Pty Ltd!",
    "  Tech  Solutions  (AU)  ",
]
```

```
for raw in inputs:
    print(f"  {raw!r:<30} -> {normalise(raw)!r}")
```

Data Processing Pipeline with reduce()

functools.reduce(fn, iterable, initial) applies fn cumulatively: result = fn(fn(fn(initial, item1), item2), item3). When fn is a pipeline step applier, reduce() threads data through an ordered list of transformations. This is more flexible than compose() because the list of steps can be built dynamically at runtime.

Example 2 — Variation

A dynamically configured processing pipeline using functools.reduce.

```
from functools import reduce

def apply_step(data: list[dict], step: Callable[[list[dict]], list[dict]]) ->
list[dict]:
    """Apply one pipeline step to a dataset."""
    return step(data)

def run_pipeline(data: list[dict], steps: list[Callable]) -> list[dict]:
    """Thread data through a list of transformation steps."""
    return reduce(apply_step, steps, data)

# Define reusable pipeline steps as named functions
def active_only(records):
    return [r for r in records if r.get("active", True)]

def add_gst(records):
    return [{**r, "total": round(r["price"] * 1.10, 2)} for r in records]

def sort_by_total(records):
    return sorted(records, key=lambda r: r["total"], reverse=True)

def top_five(records):
    return records[:5]

products = [
    {"sku": "P-001", "price": 12.50,  "active": True},
    {"sku": "P-002", "price": 89.99,  "active": False},
    {"sku": "P-003", "price": 34.95,  "active": True},
    {"sku": "P-004", "price": 149.00, "active": True},
]

result = run_pipeline(products, [active_only, add_gst, sort_by_total,
top_five])
for r in result:
    print(f"  {r['sku']}: ${r['total']:.2f}")
```

15.5 Structuring Programs with Functions

As programs grow beyond a few dozen lines, the way functions are organised becomes as important as how they are written. Two principles guide good functional structure: single responsibility (each function does one well-defined thing) and the separation of pure logic from side effects.

A pure function is one whose output depends only on its inputs and which produces no side effects — it reads no global state, writes nothing to disk or network, modifies no argument. Pure functions are trivially testable, easy to reason about, and safely composable. The goal is not to eliminate side effects (every useful program has them) but to push them to the edges of the program, keeping the core logic pure.

Single Responsibility and Layered Design

Example 1 — Introduction

A report generator structured in layers: I/O (impure) → transform (pure) → format (pure) → render (impure).

```
# IMPURE: reads from the data source (side effect)
def fetch_transactions(region: str) -> list[dict]:
    """Load transactions for a region. Actual impl would call a DB/API."""
    return [
        {"id": "T-001", "amount": 1_249.00, "status": "completed"},
        {"id": "T-002", "amount":    89.99, "status": "failed"},
        {"id": "T-003", "amount":   320.00, "status": "completed"},
    ]

# PURE: transform -- no side effects, deterministic
def filter_completed(transactions: list[dict]) -> list[dict]:
    return [t for t in transactions if t["status"] == "completed"]

def compute_summary(transactions: list[dict]) -> dict:
    amounts = [t["amount"] for t in transactions]
    return {
        "count": len(amounts),
        "total": sum(amounts),
        "mean":  sum(amounts) / len(amounts) if amounts else 0,
    }

# PURE: format -- no side effects
def format_summary(region: str, summary: dict) -> str:
    return (
        f"Region: {region}\n"
        f"  Transactions: {summary['count']}\n"
        f"  Total:        ${summary['total']:,.2f}\n"
        f"  Average:      ${summary['mean']:,.2f}"
    )

# IMPURE: renders (prints -- side effect)
def render_report(region: str) -> None:
    raw      = fetch_transactions(region)
    complete = filter_completed(raw)
    summary  = compute_summary(complete)
    print(format_summary(region, summary))

render_report("AU")
```

The main() Function and Entry Points

A Python script executed directly should define a main() function as its entry point. The if __name__ == "__main__": guard ensures that main() runs only when the file is executed directly, not when it is imported as a module by another file. This is the standard pattern for all Python scripts.

Example 2 — Variation

A fully structured command-line tool with main(), helper functions, and the __name__ guard.

```
import sys

# Pure helper functions
def parse_args(args: list[str]) -> dict:
    """Parse command-line arguments into a config dict."""
    config = {"region": "AU", "format": "text", "limit": 10}
    for arg in args[1:]:
        if "=" in arg:
            key, _, val = arg.partition("=")
            config[key.lstrip("-")] = val
    return config

def validate_config(config: dict) -> list[str]:
    """Return list of validation errors, or [] if config is valid."""
    errors = []
    if config["region"] not in {"AU", "NZ", "SG"}:
        errors.append(f"Unknown region: {config['region']!r}")
    try:
        config["limit"] = int(config["limit"])
    except ValueError:
        errors.append(f"limit must be an integer, got: {config['limit']!r}")
    return errors

# Impure main: orchestrates I/O and calls pure helpers
def main() -> int:
    """Entry point. Returns exit code."""
    config = parse_args(sys.argv)
    errors = validate_config(config)
    if errors:
        for e in errors:
            print(f"Error: {e}", file=sys.stderr)
        return 1
    print(f"Running report: region={config['region']}
limit={config['limit']}")
    return 0

if __name__ == "__main__":
    sys.exit(main())
```

15.6 Recursion with Multiple Parameters

Many practical recursive algorithms require more than one parameter: a current state, an accumulator, a bound, or a visited-nodes tracker. Tail-call style recursion — where the recursive call's result is returned directly, with no further operation — can often be made iterative for better performance. Python does not optimise tail calls, so for large inputs iterative equivalents are preferred.

GCD and Power Algorithms

Example 1 — Introduction

Euclid's GCD algorithm and fast integer exponentiation using multiple parameters.

```
def gcd(a: int, b: int) -> int:
    """
    Greatest common divisor via Euclid's algorithm.
    Works with any two non-negative integers.
    """
    if b == 0:
        return a
    return gcd(b, a % b)

def lcm(a: int, b: int) -> int:
    """Least common multiple using GCD."""
    return abs(a * b) // gcd(a, b)

# Fast exponentiation: O(log n) instead of O(n)
def power(base: int, exp: int, mod: int = None) -> int:
    """
    Compute base**exp, optionally modulo mod.
    Uses divide-and-conquer: O(log exp) multiplications.
    """
    if exp == 0:
        return 1
    if exp % 2 == 0:
        half = power(base, exp // 2, mod)
        result = half * half
    else:
        result = base * power(base, exp - 1, mod)
    return result % mod if mod else result

print(f"GCD(48, 18) = {gcd(48, 18)}")          # 6
print(f"LCM(12, 18) = {lcm(12, 18)}")          # 36
print(f"3**10 = {power(3, 10)}")                # 59049
print(f"3**10 mod 1000 = {power(3, 10, 1000)}")  # 49
```

Tree Traversal and Accumulation

Example 2 — Variation

Recursive tree traversal with path accumulation: finding all leaf nodes and their full paths.

```
def find_leaf_paths(
    node:    dict,
    current_path: list[str] = None,
    results: list[tuple]    = None,
) -> list[tuple[list[str], object]]:
    """
    Traverse a nested dict tree and return (path, value) for every leaf.
    A leaf is any value that is not a dict.
    """
    if current_path is None: current_path = []
```

```
    if results is None:      results = []

    for key, value in node.items():
        path = current_path + [key]
        if isinstance(value, dict):
            find_leaf_paths(value, path, results)  # recurse
        else:
            results.append((path, value))          # leaf: record

    return results

config = {
    "database": {"host": "db-01", "port": 5432, "pool": {"size": 10,
"timeout": 30}},
    "api":      {"host": "api-01", "port": 443},
    "debug":    False,
}

for path, value in find_leaf_paths(config):
    dotted = ".".join(path)
    print(f"  {dotted:<35} = {value}")
```

15.7 Introduction to Methods: Functions Inside Classes

A method is a function defined inside a class body. When called on an object, Python automatically passes the object as the first argument (conventionally named self). This is the bridge from functions to object-oriented programming: everything covered in this book about writing good functions applies equally to methods. The concepts are identical — only the context changes.

Instance Methods

An instance method receives the object (self) as its first parameter and has access to all of the object's attributes. Type hints, docstrings, *args, **kwargs, and all other function features apply to methods. The __init__ method is the constructor: it initialises the object's state when the class is instantiated.

Example 1 — Introduction

A ServiceMonitor class where each method is a well-typed, documented function.

```
from datetime import datetime
from typing import Optional

class ServiceMonitor:
    """Tracks health metrics for a single service."""

    def __init__(self, service_name: str, alert_threshold: float = 90.0) ->
None:
        """
        Args:
            service_name:     Name of the service being monitored.
            alert_threshold:  CPU % above which an alert is triggered.
        """
        self.service_name     = service_name
```

```
        self.alert_threshold  = alert_threshold
        self._readings: list[dict] = []

    def record(self, cpu_pct: float, latency_ms: int) -> None:
        """Record a new metrics sample."""
        self._readings.append({
            "timestamp": datetime.now().isoformat(timespec="seconds"),
            "cpu_pct":   cpu_pct,
            "latency_ms": latency_ms,
        })

    def average_cpu(self) -> Optional[float]:
        """Return the mean CPU % across all readings, or None if no data."""
        if not self._readings:
            return None
        return sum(r["cpu_pct"] for r in self._readings) / len(self._readings)

    def is_healthy(self) -> bool:
        """Return True if no reading exceeds the alert threshold."""
        return all(r["cpu_pct"] < self.alert_threshold for r in
self._readings)

    def summary(self) -> str:
        """Return a human-readable summary of current health."""
        avg = self.average_cpu()
        status = "HEALTHY" if self.is_healthy() else "DEGRADED"
        avg_str = f"{avg:.1f}%" if avg is not None else "no data"
        return f"{self.service_name}: [{status}] avg CPU={avg_str}"

monitor = ServiceMonitor("api-gateway", alert_threshold=85.0)
for cpu, latency in [(62, 45), (71, 52), (88, 920), (65, 48)]:",
    monitor.record(cpu, latency)

print(monitor.summary())
print(f"Average CPU: {monitor.average_cpu():.1f}%")
```

@classmethod and @staticmethod

@classmethod receives the class itself (cls) as the first argument rather than an instance. It is used for alternative constructors and factory methods. @staticmethod receives neither self nor cls: it is a regular function that happens to live in the class namespace because it is logically related to the class.

Example 2 — Variation

@classmethod as an alternative constructor; @staticmethod for a utility calculation.

```
class MonitorConfig:
    """Configuration for a service monitor."""

    DEFAULT_THRESHOLD = 90.0

    def __init__(self, service: str, threshold: float, interval_s: int) ->
None:
        self.service    = service
        self.threshold  = threshold
```

```
        self.interval_s = interval_s

    @classmethod
    def from_dict(cls, data: dict) -> "MonitorConfig":
        """Alternative constructor: build from a config dictionary."""
        return cls(
            service    = data["service"],
            threshold  = data.get("threshold", cls.DEFAULT_THRESHOLD),
            interval_s = data.get("interval_s", 30),
        )

    @staticmethod
    def samples_per_hour(interval_s: int) -> int:
        """Calculate how many samples fit in one hour."""
        return 3_600 // interval_s

    def __repr__(self) -> str:
        return (f"MonitorConfig(service={self.service!r}, "
                f"threshold={self.threshold}, interval_s={self.interval_s})")

# Regular constructor
cfg1 = MonitorConfig("auth", 85.0, 15)

# Alternative constructor via @classmethod
cfg2 = MonitorConfig.from_dict({"service": "db", "interval_s": 60})

# @staticmethod: called on class, no instance needed
print(f"Samples/hour at 15s: {MonitorConfig.samples_per_hour(15)}")
print(cfg1)
print(cfg2)
```

Chapter 15 Summary

- Functions are first-class objects in Python: they can be stored in variables and data structures, passed as arguments to other functions, and returned as values. A function that receives or returns another function is a higher-order function. Dispatch tables (dicts of functions), callbacks, and factory functions all follow from this property.
- A closure is an inner function that captures variables from its enclosing scope. Closures provide private, persistent state without classes. Factory functions return closures configured with specific parameters. Common patterns: rate limiters, specialised validators, and formatters built via closures.
- A generator function contains yield and returns a generator object. Generators produce values lazily on demand, making them memory-efficient for large or infinite sequences. Multiple generators can be chained into O(1)-memory pipelines. Use itertools.islice() to safely consume a bounded slice of an infinite generator.
- Function composition combines multiple single-purpose functions into one. compose(*fns) applies functions left-to-right. functools.reduce() can thread data through a dynamic list of pipeline steps. Both enable building complex operations from small, testable, reusable parts.
- Structure programs in layers: pure logic functions (deterministic, no side effects, easy to test) in the interior; impure I/O functions (reading files, calling APIs, printing) at the edges. The main()

function orchestrates the program and is guarded by if __name__ == "__main__":, which prevents execution when the module is imported.

- Multi-parameter recursion: each recursive call reduces the problem by updating two or more parameters simultaneously (GCD, fast power, tree traversal with path accumulator). Avoid mutating mutable default arguments as accumulator state (use None and initialise inside). For very deep recursion, prefer an iterative approach.
- A method is a function inside a class body. Instance methods receive self automatically. __init__ initialises object state. @classmethod receives the class (cls) and is used for alternative constructors. @staticmethod receives neither and is a utility function logically grouped with the class. All function principles — type hints, docstrings, *args, **kwargs — apply equally to methods.

Guided Exercises

Complete the following exercises in VS Code, IDLE, or your preferred Python environment. Each exercise targets a specific concept from this chapter.

1. Build a function transformation registry. Create a TRANSFORMERS dict mapping operation names to functions (at least ten operations: strip, lower, upper, to_int, to_float, to_bool, truncate_N, pad_left_N, replace_X_with_Y, and one of your own design). Write an apply_transforms(value, *operations) function that applies a sequence of named operations. Handle unknown operations with a descriptive error. Demonstrate with at least five different transformation chains on realistic data.

2. Implement a generator-based log analysis pipeline. Write at least four generator functions: read_lines(source) yields lines from a list; parse_record(lines) parses each line into a dict (skipping malformed lines); filter_by_level(records, min_level) yields only records at or above the level; and enrich_record(records) adds a parsed_timestamp field. Chain all four into a pipeline and produce a summary: total records processed, count by level, and first and last timestamps.

3. Write a make_validator() factory function that returns a configured validator. The factory should accept: required fields (set of names), type constraints (dict of field->type), range constraints (dict of field->(min, max)), and custom validators (dict of field->callable). The returned validator function should accept a dict and return (is_valid: bool, errors: list[str]). Demonstrate with three different validator instances for different record types (product, server config, user account), testing valid and invalid inputs.

4. Implement a three-stage generator pipeline for processing a simulated event stream. Stage 1: generate_events() yields dicts with timestamp, type, service, and payload. Stage 2: enrich_events() adds a severity field based on event type. Stage 3: throttle(events, rate_per_second) yields events but inserts delays to enforce the rate limit. Combine the three stages and consume the first 20 events, printing a formatted summary of each.

5. Design a ServiceMonitor class (as introduced in 15.7) with full implementation. The class should maintain a rolling window of readings (configurable size, using collections.deque). Add methods: record(cpu_pct, latency_ms), average_cpu(), peak_cpu(), is_healthy(), trend() (returning "improving", "stable", or "degrading" based on recent readings), and a summary() method. Add @classmethod from_config(cls, config_dict) and @staticmethod validate_threshold(threshold). Write a main() function that creates three monitors and simulates 20 readings each.

Challenge Tasks

The following tasks are open-ended. No hints are provided. They require you to design a solution, not just apply a pattern.

1. Design and implement a fully functional event-driven processing system using generators and closures. The system should: produce events from a simulated source using a generator; route events to handlers using a dispatch dict; allow handlers to be registered and deregistered dynamically using closure-based factories; collect metrics (events processed, errors, processing time) using a closure-based accumulator; and generate a report after processing 100 events. Every component should be a pure function or a generator; all state should be managed through closures or function arguments rather than global variables.

2. Build a complete lazy data pipeline framework. Implement a Pipeline class that wraps a generator and supports method chaining: .filter(predicate), .map(transform), .flat_map(transform), .take(n), .skip(n), .batch(size) (yield lists of N elements), .throttle(rate_per_second), .with_retry(max_attempts, exceptions), and .collect() (materialise to list). Each method should return a new Pipeline wrapping a new generator. Demonstrate with a pipeline that reads a simulated stream of at least 1000 records, filters, transforms, batches, and collects results, printing memory and timing statistics.

3. Implement a dependency injection container using closures and first-class functions. The container should: register services by name, either as instances or as factory functions (registered with a lifetime of "singleton" or "transient"); resolve services by name, automatically resolving their constructor dependencies (inspect function signatures using the inspect module); detect circular dependencies and raise a descriptive error; and support scoped resolution where a subset of registrations is overridden within a context. Write at least six service classes with non-trivial dependency graphs and demonstrate resolution of all services, singleton lifetime verification, and circular dependency detection.

APPENDIX A How to Create and Use Flow Diagrams Before Coding

A flow diagram (flowchart) maps the logical flow of a program before you write a single line of code. It forces you to think through every branch, loop, and edge case in plain visual terms. Programs built from a clear diagram tend to have fewer structural bugs and are easier to test, explain, and extend.

Flowchart Symbols

Four symbols cover the vast majority of program logic:

- Oval / rounded rectangle — Terminator (Start / End)
- Rectangle — Process: an action or calculation
- Diamond — Decision: a yes/no or true/false branch
- Arrow — Flow: direction of execution between symbols

A 10-Step Process for Any Program

1. State the problem in one sentence. What does the program do?
2. Identify the inputs. What data does the program receive?
3. Identify the outputs. What results does the program produce?
4. Identify the decisions. What conditions change the program's path?
5. Identify loops. What actions repeat, and what stops them?
6. Sketch the flowchart on paper or in a tool (Lucidchart, draw.io, Mermaid).
7. Trace through each path manually with sample inputs.
8. Check all branches terminate correctly (no infinite paths).
9. Translate to pseudocode (Appendix C).
10. Convert pseudocode to Python, testing one block at a time.

Example 1: Connection Retry Loop

Problem: Attempt to connect to a database. Retry up to MAX_RETRIES times with a delay between attempts. If all attempts fail, log an error and exit.

Flow: Start → Set attempts = 0 → [attempts < MAX_RETRIES?] — No → Log failure, End — Yes → Attempt connection → [Connected?] — Yes → Proceed, End — No → Increment attempts, Wait delay → back to loop condition

```
# Python from the flowchart above
import time
```

```
MAX_RETRIES  = 5
RETRY_DELAY  = 2.0

attempts  = 0
connected = False

while attempts < MAX_RETRIES and not connected:
    attempts += 1
    connected = attempt_connection()   # returns True on success
    if not connected and attempts < MAX_RETRIES:
        time.sleep(RETRY_DELAY)

if connected:
    print(f"Connected after {attempts} attempt(s)")
else:
    print("All attempts failed")
```

Example 2: Data Validation Gate

Problem: Accept a record only if all three fields pass validation (customer_id non-empty, amount positive float, currency in approved list). Collect all errors before rejecting.

Flow: Start → Initialise errors = [] → Check customer_id [empty?] Yes → Append error → Check amount [parseable & positive?] No → Append error → Check currency [in approved set?] No → Append error → [errors empty?] Yes → Process record — No → Report errors, End. Each validation is an independent diamond; all three are always evaluated.

Using Flowcharts for Larger Programs

For programs with multiple functions, draw one high-level flowchart showing function relationships, then a detail flowchart for any function with non-trivial logic. Tools that integrate with code (Mermaid in Markdown, PlantUML) let diagrams live next to the source and stay in sync as the code evolves.

APPENDIX B Suggested Projects for Continued Practice

The following projects are arranged in three tiers: Foundational (chapters 1–10 skills), Intermediate (chapters 11–13), and Advanced (chapters 14–15). Each specification lists the primary concepts it exercises and a suggested minimum scope.

Tier 1: Foundational Projects (Chapters 1–10)

1. Unit Conversion Library (3–4 days)

Build a command-line unit converter supporting at least three categories: length (mm, cm, m, km, in, ft, miles), mass (g, kg, oz, lb, tonnes), and temperature (Celsius, Fahrenheit, Kelvin). Use a menu loop, input validation with try/except, and f-string formatted output. Primary concepts: input/output, arithmetic, conditional statements, while loops.

2. Expense Tracker (4–5 days)

Record income and expense entries (description, amount, category) during a session. Compute totals, balance, largest expense, and a breakdown by category. Use a dispatch-table menu loop with a sentinel exit command. Primary concepts: lists, dicts, string formatting, loop control, type hints.

3. Password Strength Checker (3 days)

Evaluate a password against at least six rules: minimum length, uppercase, lowercase, digits, special characters, and absence of common patterns. Report each failing rule, then add a password generator. Primary concepts: strings, loops, conditions, sets for character classification.

4. CSV Data Cleaner (5 days)

Read a CSV file with messy data (extra whitespace, inconsistent casing, empty fields, invalid numeric values). Apply a configurable cleaning pipeline: strip, normalise case, fill missing values, remove invalid rows. Write cleaned output to a new file. Primary concepts: string methods, lists, dicts, file I/O, error handling.

Tier 2: Intermediate Projects (Chapters 11–13)

5. Contact Directory (5–6 days)

A contact manager storing name, email, phone, and tags (as a frozenset) per contact. Support add, search by name or email, update, tag-based filtering, and CSV export. Index contacts by email with a dict for O(1) lookup. Demonstrate tag intersection/union queries. Primary concepts: lists of dicts, dict methods, sets, namedtuples.

6. Inventory Management System (6–7 days)

Manage a product catalogue (SKU, name, category, price, stock) with a transaction log. Support: receive stock, dispatch stock (prevent negative), search, low-stock report, and category summary. Use a dict indexed by SKU and defaultdict for grouping. Primary concepts: dicts, defaultdict, Counter, list comprehensions, sorting.

7. Service Health Monitor (5 days)

Simulate a monitoring dashboard. Maintain service records as namedtuples (name, host, port, status, cpu_pct, error_rate). Implement health classification, fleet-wide statistics, and change detection between two snapshots using set operations. Primary concepts: namedtuples, sets, frozensets, boolean logic.

8. Log File Analyser (6–7 days)

Parse a structured log file (timestamp, level, service, message). Group by service using defaultdict(list), count by level using Counter, find first/last occurrence of each error code, and produce a formatted report. Primary concepts: string methods, dicts, Counter, sorted output.

Tier 3: Advanced Projects (Chapters 14–15)

9. Typed Configuration System (5–6 days)

Build a configuration loader that reads key=value files, validates each field against a schema (type, required, range, custom validator), applies type conversion, and returns a typed config dict. Type hints throughout. Primary concepts: functions, type hints, closures for validators, try/except, docstrings.

10. Data Processing Pipeline (7–8 days)

Build a lazy pipeline using generator functions. Stages: read, parse and type-convert, filter, enrich, batch, write. Each stage is a generator; the pipeline processes records one at a time (O(1) memory). Wrap stages with timing and logging decorators. Primary concepts: generators, closures, decorators, functools, first-class functions.

11. Task Queue Processor (8–10 days)

A priority job queue where jobs have id, priority (1–5), payload, status, and attempt count. Process highest-priority pending jobs first. Re-queue failed jobs up to MAX_ATTEMPTS. Use a dispatch table (dict of status → handler function) to route outcomes. Primary concepts: functions, closures, dispatch tables, namedtuples, sorted().

12. Portfolio Project: Automated Report Generator (2–3 weeks)

A fully documented report generator. Input: a dataset (CSV or list of dicts) and a report specification (dict defining filters, groupings, aggregations, sort order, output format). Use type hints on all functions, docstrings, a generator-based data pipeline, a timing decorator, defaultdict for aggregation, Counter for frequency fields, and the main() pattern with if __name__ == "__main__":. Include a README.

APPENDIX C Pseudocode – Purpose, Usage, and Conversion

1. What is Pseudocode?

Pseudocode is a structured informal description of an algorithm. It uses the logical constructs of a programming language (sequence, selection, iteration) while ignoring language-specific syntax. Its purpose is to let the designer focus on what the program must do before deciding how to express it in Python.

Pseudocode is not executed. It lives in the planning phase and serves as a bridge between the flow diagram (visual logic) and the Python source code (formal syntax).

2. Why Use Pseudocode?

Writing pseudocode before code produces fewer structural bugs, cleaner function boundaries, and code that is easier to read and maintain. It forces you to resolve every branch and edge case at low cost — editing plain text is far faster than refactoring code. Pseudocode is also the shared language for design discussions with stakeholders who do not know Python.

3. Characteristics of Good Pseudocode

- Capitalised keywords: BEGIN, END, IF, THEN, ELSE, WHILE, FOR, REPEAT, UNTIL, RETURN
- Plain English for actions: "Prompt user for port number", not raw Python syntax
- Consistent indentation to show block structure
- Descriptive variable names: customer_id not c
- No language-specific syntax (no colons, parentheses, import statements)
- Detailed enough to translate directly to code, but no more

4. Example 1: Input Validation Loop

Problem: Prompt the user for a port number. Accept only integers in the range 1–65535. Repeat until valid input is received.

```
BEGIN
  REPEAT
    PROMPT user "Enter port number (1-65535): "
    READ raw_input
    IF raw_input is not a valid integer THEN
      DISPLAY "Please enter a whole number"
    ELSE IF integer value < 1 OR integer value > 65535 THEN
      DISPLAY "Port must be between 1 and 65535"
    ELSE
      SET port = integer value
      EXIT loop
    END IF
```

```
  UNTIL port is set
  RETURN port
END
```

Equivalent Python:

```
def prompt_port() -> int:
    while True:
        raw = input("Enter port number (1-65535): ")
        try:
            port = int(raw)
        except ValueError:
            print("Please enter a whole number")
            continue
        if not (1 <= port <= 65535):
            print("Port must be between 1 and 65535")
        else:
            return port
```

5. Example 2: Data Grouping Algorithm

Problem: Group a list of transaction records by region and compute the total amount of completed transactions per region.

```
BEGIN
  SET grouped = empty dictionary

  FOR each transaction in transactions
    IF transaction.status equals "completed" THEN
      IF transaction.region NOT IN grouped THEN
        SET grouped[transaction.region] = 0
      END IF
      ADD transaction.amount TO grouped[transaction.region]
    END IF
  END FOR

  FOR each region, total IN grouped
    DISPLAY region + ": $" + total (2 decimal places)
  END FOR
END
```

Equivalent Python:

```
from collections import defaultdict

grouped = defaultdict(float)
for txn in transactions:
    if txn["status"] == "completed":
        grouped[txn["region"]] += txn["amount"]

for region, total in sorted(grouped.items()):
    print(f"{region}: ${total:,.2f}")
```

6. Pseudocode to Python Translation Table

BEGIN / END → function body or module-level code

PROMPT / READ → input() with appropriate type conversion

DISPLAY → print()

IF / THEN / ELSE / END IF → if / else (indented block)

WHILE condition / END WHILE → while condition:

FOR each item IN collection → for item in collection:

REPEAT / UNTIL condition → while True: ... if condition: break

RETURN value → return value

SET variable = value → variable = value

APPEND item TO list → list.append(item)

ADD value TO total → total += value

APPENDIX D Project Demonstration – Expense Tracker

This appendix walks through the complete development lifecycle of a command-line Expense Tracker: problem definition, user flow, pseudocode, Python implementation, and documentation. It demonstrates the practices taught throughout the book: type hints, docstrings, validated input, modular functions, a dispatch-table menu, Optional typing, and the main() pattern.

Step 1: Define the Project

Objective:

A command-line program that records income and expense entries during a session, computes totals and balance, reports the largest expense, and lists entries filtered by category.

Features:

- Add an income or expense entry (description, amount, category)
- View a running summary (total income, total expenses, balance, largest expense)
- View all entries
- View entries filtered by category
- Exit cleanly

Constraints:

Terminal only, all data in memory. Amount must be a positive float; category must be a non-empty string. Invalid input is rejected with a specific error message — the program must never crash on bad input.

Step 2: User Flow

[Start] → [Initialise entries list] → [Display menu] → [Read choice]

→ [1] Add income: prompt description, amount, category → validate all → append → back to menu

→ [2] Add expense: same as income → back to menu

→ [3] Show summary: compute & print totals and balance → back to menu

→ [4] View all: print each entry → back to menu

→ [5] Filter by category: prompt category → print matching entries → back to menu

→ [Q] Exit: print goodbye → [End]

Step 3: Pseudocode

```
BEGIN
  INITIALISE entries as empty list

  WHILE user has not chosen to quit
    DISPLAY menu
    READ choice

    IF choice is 1 or 2 THEN
      PROMPT description (non-empty string)
      PROMPT amount (positive float)
      PROMPT category (non-empty string)
      APPEND {type, description, amount, category} to entries

    ELSE IF choice is 3 THEN
      COMPUTE total income, total expenses, balance
      DISPLAY totals, balance, largest expense

    ELSE IF choice is 4 THEN
      FOR each entry in entries: DISPLAY formatted entry

    ELSE IF choice is 5 THEN
      PROMPT category name
      FOR each entry matching category: DISPLAY entry

    ELSE IF choice is Q THEN
      EXIT loop

    ELSE
      DISPLAY "Unrecognised option"
    END IF
  END WHILE
  DISPLAY goodbye message
END
```

Step 4: Python Implementation

```
import sys
from typing import Optional

Entry = dict[str, object]


def prompt_nonempty(prompt: str) -> str:
    """Prompt until a non-empty string is entered."""
    while True:
        val = input(prompt).strip()
        if val:
            return val
        print("  This field cannot be empty.")


def prompt_positive_float(prompt: str) -> float:
    """Prompt until a positive float is entered."""
    while True:
        raw = input(prompt).strip()
        try:
            value = float(raw)
            if value > 0:
                return value
            print("  Amount must be greater than zero.")
        except ValueError:
            print(f"  Invalid amount: {raw!r}")


def add_entry(entries: list[Entry], entry_type: str) -> None:
    """Collect and append a new income or expense entry."""
    description = prompt_nonempty("  Description: ")
    amount      = prompt_positive_float("  Amount ($): ")
    category    = prompt_nonempty("  Category:    ")
    entries.append({
        "type":        entry_type,
        "description": description,
        "amount":      amount,
        "category":    category.lower(),
    })
    print(f"  Added: {description} ${amount:.2f} [{category}]")


def show_summary(entries: list[Entry]) -> None:
    """Print totals, balance, and largest expense."""
    income   = sum(e["amount"] for e in entries if e["type"] == "income")
    expenses = sum(e["amount"] for e in entries if e["type"] == "expense")
    exp_list = [e for e in entries if e["type"] == "expense"]
    largest  = max(exp_list, key=lambda e: e["amount"]) if exp_list else None
    print(f"\n  Income:   ${income:>10,.2f}")
    print(f"  Expenses: ${expenses:>10,.2f}")
    print(f"  Balance:  ${income - expenses:>10,.2f}")
    if largest:
        print(f"  Largest:  {largest['description']}
${largest['amount']:.2f}")
```

```
def list_entries(
    entries:  list[Entry],
    category: Optional[str] = None,
) -> None:
    """Print all entries, optionally filtered by category."""
    filtered = [
        e for e in entries
        if category is None or e["category"] == category.lower()
    ]
    if not filtered:
        print("  No entries found.")
        return
    for e in filtered:
        sign = "+" if e["type"] == "income" else "-"
        print(f"  [{e['category']:<12}] {e['description']:<25}
{sign}${e['amount']:,.2f}")


def show_menu() -> None:
    print("\n=== Expense Tracker ===")
    print("[1] Add income    [2] Add expense")
    print("[3] Summary       [4] View all")
    print("[5] Filter by category    [Q] Quit")


def main() -> int:
    """Run the tracker. Returns exit code."""
    entries: list[Entry] = []
    DISPATCH = {
        "1": lambda: add_entry(entries, "income"),
        "2": lambda: add_entry(entries, "expense"),
        "3": lambda: show_summary(entries),
        "4": lambda: list_entries(entries),
    }
    while True:
        show_menu()
        choice = input("> ").strip().lower()
        if choice in DISPATCH:
            DISPATCH[choice]()
        elif choice == "5":
            cat = prompt_nonempty("Filter by category: ")
            list_entries(entries, cat)
        elif choice == "q":
            print("Goodbye.")
            return 0
        else:
            print("  Unrecognised option.")


if __name__ == "__main__":
    sys.exit(main())
```

Step 5: Documentation Checklist

Before submission or sharing:

- Every function has type hints on all parameters and the return value
- Every function has a docstring
- Input validation never lets an invalid value reach the business logic
- main() is called only via if __name__ == "__main__":
- All variable names are descriptive and in snake_case
- No global state — all data is passed as arguments

Possible extensions:

- Persist entries across sessions using json.dump / json.load
- Export to CSV with the csv module
- Category breakdown report using Counter
- Add a @timed decorator to measure each menu handler

APPENDIX E Glossary of Terms

Abstraction: Hiding complex implementation details and exposing only the essential interface. Functions and classes are both forms of abstraction.

Algorithm: A finite, ordered set of steps that solves a problem or performs a computation.

Argument: A value supplied to a function at the point of call. Distinguished from parameter, which is the name in the function definition.

Boolean: A value that is exactly True or False. In Python, bool is a subclass of int (True == 1, False == 0).

Callable: Any object that can be called with parentheses: functions, lambdas, classes, and objects defining __call__.

Casting: Explicitly converting a value from one type to another using int(), float(), str(), or similar constructors.

Class: A blueprint for creating objects. Defines the attributes (data) and methods (functions) that instances will have.

Closure: An inner function that captures and remembers variables from its enclosing scope, even after the outer function has returned.

Comment: Text in source code beginning with # that is ignored by the interpreter. Used to explain intent.

Comprehension: A concise expression for creating a list, set, or dict from an iterable: [x*2 for x in items if x > 0].

Conditional Statement: An if / elif / else construct that executes different code depending on whether conditions are True or False.

Constructor: The __init__ method of a class, called automatically when an object is created with ClassName().

Counter: A dict subclass from collections that counts element occurrences. Supports most_common(n) and arithmetic.

Decorator: A function that wraps another function to add behaviour before or after the call. Applied with @decorator_name.

defaultdict: A dict subclass from collections that calls a factory function for missing keys instead of raising KeyError.

Dictionary: An ordered (Python 3.7+), mutable mapping of unique immutable keys to values, created with {} or dict().

Docstring: A string literal immediately inside a function, class, or module body. Accessible via __doc__ and help().

Exception: An object representing an error. Caught with try/except; raised with raise. Allows programs to handle errors gracefully.

f-string: A formatted string literal (Python 3.6+) prefixed with f. Expressions inside {} are evaluated at runtime.

frozenset: An immutable variant of set. Hashable and usable as a dictionary key or as an element of another set.

Function: A named, reusable block of code defined with def. Accepts parameters, performs a task, and optionally returns a value.

Generator: A function containing yield that returns a generator object. Produces values lazily, one at a time, on demand.

Global Variable: A variable defined at module level. Avoid modifying with the global keyword; prefer passing values as arguments.

IDE: Integrated Development Environment. Software combining a code editor, debugger, and run environment. Examples: VS Code, PyCharm.

Immutable: An object whose value cannot be changed after creation. Strings, tuples, integers, and frozensets are immutable.

Index: An integer position within a sequence. Python uses zero-based indexing; negative indices count from the end.

Inheritance: A mechanism where a class (subclass) acquires attributes and methods from another class (superclass).

Input: Data received by a program, typically via input(), a file, a network socket, or command-line arguments.

Iterable: Any object that can be looped over: lists, tuples, strings, dicts, sets, generators, and any object with __iter__.

Iteration: Repeatedly executing a block of code for each element in an iterable, or while a condition holds.

Keyword Argument: An argument passed by name: fn(timeout=30). Can appear in any order after all positional arguments.

Lambda: An anonymous, single-expression function: lambda x, y: x + y. Used as a concise key= or callback argument.

LEGB Rule: The name-resolution order Python uses: Local → Enclosing → Global → Built-in.

List: An ordered, mutable sequence created with []. Supports indexing, slicing, appending, removing, and sorting.

Local Variable: A variable defined inside a function. Exists only for the duration of that function call.

Logical Operator: and, or, and not — operators that combine boolean values. and and or use short-circuit evaluation.

Loop: A construct that repeats a block of code: for (known iterations over an iterable) or while (condition-driven).

lru_cache: functools.lru_cache — a decorator that memoises function results keyed by arguments. Avoids redundant computation.

Method: A function defined inside a class. Instance methods receive self; class methods receive cls; static methods receive neither.

Module: A Python file containing definitions and statements. Imported with import to make its contents available.

Mutable: An object whose value can be changed in place: lists, dicts, and sets are mutable.

namedtuple: A tuple subclass from collections with named fields. Combines tuple immutability with readable attribute access.

Nested: A data structure or control construct placed inside another of the same kind: nested lists, nested loops, nested functions.

None: Python's null singleton. Returned by functions with no explicit return statement. Always test with is None.

Object: An instance of a class. Every Python value — including functions and classes themselves — is an object.

Output: Data produced by a program, typically via print(), a file, or a network response.

Parameter: A variable in a function definition that receives an argument value when the function is called.

Pseudocode: An informal, structured description of an algorithm using English and control-flow keywords. Not executable.

Pure Function: A function whose output depends only on its inputs, with no side effects. Deterministic and easy to test.

Recursion: A function calling itself. Requires a base case to terminate and a recursive case to reduce the problem.

Return Statement: return value sends a value back from a function to its caller. No return statement returns None.

Runtime Error: An error that occurs during execution, such as ZeroDivisionError, KeyError, or TypeError.

Scope: The region of a program where a name is visible. Resolved by the LEGB rule.

Sentinel Value: A special value that signals end-of-input in a loop, causing the loop to stop processing.

Set: An unordered collection of unique hashable elements with O(1) membership testing. Supports union, intersection, difference.

Slice: A portion of a sequence extracted with [start:stop:step]. Returns a new sequence without modifying the original.

String: An immutable ordered sequence of Unicode characters. Supports indexing, slicing, and a rich library of methods.

Syntax Error: An error caused by code that violates Python's grammar rules. Detected before execution begins.

Tuple: An immutable ordered sequence created with (). Hashable when all elements are hashable; usable as a dict key.

Type Hint: An optional annotation (Python 3.5+) on a parameter or return value: def f(x: int) -> str. Not enforced at runtime.

Variable: A name bound to a value. Assignment (=) creates or rebinds the name; the underlying value may or may not change.

While Loop: A loop that continues while a condition is True. Requires initialisation, condition, and update to avoid infinite loops.

INDEX

Think, Code, Create – A Python Programming Journey

References are to chapters (Ch.) and appendices (App.).

Made in the USA
Monee, IL
15 September 2026

63228642R00142